Praise for Danica McKellar and *Kiss My Math*

"McKellar is probably the only person on prime-time television who moonlights as a cyberspace math tutor."   —The New York Times

"If you liked Danica McKellar's first book, *Math Doesn't Suck* . . . then you should get *Kiss My Math*. . . . It is every bit as good as the first. Besides, think how cool it is to walk into your math class with a book having that title under your arm!"
                    —Dr. Keith Devlin, Stanford University, NPR's "Math Guy" and the author of
                                The Numbers Behind NUMB3RS and The Math Instinct

"Danica McKellar has a message for girls: Cute and smart is better than cute and dumb."   —The Associated Press

"It's really hard to put the book down, it's so delightfully written. . . . The real power of the book is actually in these memorable analogies and in the very friendly tone she uses all through the book. She's talking to a friend, and it makes you just keep reading!"   —Home School Math Blogs

"One might ask, 'How is [*Kiss My Math*] any different from the math books that schools and libraries already have?' McKellar understands teens. She speaks their language. . . . And honestly, she makes math fun. Anyone who can do that has a hit on her hands."   —VOYA (Voice of Youth Advocates)

"Every page of this winning book is lively and interesting, filled with quotes from teenagers about their successes in math, zippy little drawings, and short-cuts for solving even the thorniest math problems."   —The Boston Globe

"For everyone having trouble with math, get this down-to-earth book that will pull your grades up and intrigue even the most reluctant student."
                                                —The New Mexican

"At the end of the book is a 'Math Test Survival Guide,' which may be worth the price of the book in itself."   —Let's Play Math Blogs

Kiss My Math in Action!
A Sampling of Danica's Emails

OMG, I used to be horrible at math, and soooo nervous about math tests. But since I started reading your books, I'm not so nervous anymore, and I even got 100% on my last math test! Thank you so much!!
                                                —Ella, age 13, New York

Recently we had a quiz and I was positively terrified of my grade because the class before us was walking out saying that the average was a D! Mine was the last one. I took a deep breath as the teacher handed it to me. My heart was beating so fast. I glanced at the paper where a number was circled in a red correcting pen: 95%! I smiled so hard that I almost laughed! She asked me where I knew all of the material from and I said "Class . . . and one amazing book called *Kiss My Math*. It gives really cool methods on algebra." I handed it to her, and she asked me if she could borrow it to look through! I am so proud of myself to have come that far in math. I can't believe what happened to me. You are an amazing person that can make magic. Just one request: keep writing your books! —Lucy, age 12, Massachusetts

You are my new hero. I'm on the third chapter of *Kiss My Math*, and I swear it's like you are in my head or something! I understand pre-algebra so much better, it's like a miracle. I just don't know how you did it—you have not only helped me in pre-algebra, but it has improved my outlook on the whole subject. Thank you so much for taking the time to write such awesome books to help girls throughout the country. I mean, being 12 is hard enough! I don't need pre-algebra to stress over, and now I don't have to worry about that. Again, THANK YOU THANK YOU THANK YOU. You are a lifesaver.
—Chloe, age 12, Alabama

As a homeschooled girl, the real challenge for me has been math—my worst and most hated subject—until I got your books! I quickly realized that the methods you were using weren't boring like my math textbook. By the time I finished, things I couldn't understand before were almost easy! And you've shown me that a girl can be beautiful *and* smart. I've never dumbed myself down for a guy, and I never will. Thank you! —Kenzie, age 14, Indianapolis

I love *Kiss My Math*! A few days ago I was crying while I was doing math. My mom came over and told me that I shouldn't worry and that I'll get it, and the cool thing is, I was crying because I did get it. I was just so happy! Thank you, Danica! So much! —Shanni, age 15, Ohio

You are a role model for being a star, and still being smart. I used to hate math and get Cs. Because of you, now I LOVE math and I'm getting an A–. Thank you! —Madison, age 12, Iowa

I got *Kiss My Math* a few days ago, and I just totally aced a math test because of you! I am extremely grateful. Thank you sooooo much for helping girls with math in such a fun, girl-friendly, stress-free way. —Rachael, age 13, Ohio

I've always liked math, and wanted to do my best in it, but still struggled. The problem was, I didn't know how to get better scores. When I read your book, I understood math a whole lot better and finally got into the advanced math class—the one for kids who are very smart in math and need more of a challenge. Thank you so much!   —Emily, age 13, Wisconsin

I wanted to really thank you for making math really FUN. I am really enjoying this book and I have told a bunch of my friends about it and they are all going to get it now!   —Colleen, age 12, Michigan

I love how you make math seem so easy for gals my age, and I sometimes laugh out loud when I read your books! Thanks to you, *I'm proud to be a smart girl.*   —Kira, age 12, Virginia

Your book *Kiss My Math* saved my life! I'm in algebra in the eighth grade and your book covers it all. I have hated math all my life and now I am enjoying math class!   —Kate, age 13, Missouri

I used your "Pretend Cheat Sheet" from the Math Test Survival Guide in *Kiss My Math* to study for my last math test, and I finished it early and got my second highest grade ever! Thank you!   —Annie, age 13, California

Thanks to you, I've gone from a D+ to an A in math class. It's given me a whole new confidence in myself. I also think you're a great role model, thanks so much for everything!   —Hailey, age 12, California

I have been teaching math for 30 years and love your techniques. Using your creative methods in the classroom, the kids REALLY get the math. We love your books!"   —Judy-Ann Ehrlich, Blake School, Minneapolis, Minnesota

I have been teaching math for 26 years and now I am using both of your books as a resource for my class notes and class work this year. I love the manner in which you cover the important skills. You have made my job so much easier.

—Debbie Perry, Stranahan High, Fort Lauderdale, Florida

Our son started 7th grade pre-algebra this year and, while he is an A student in other subjects, his math class brought him (and us) to our knees. *Kiss My Math* humanizes some of the more technical aspects of math, and makes the concepts so much easier to relate to and to remember. Our son just finished the first trimester with a B+ and we are certain this would have been a C or less if we did not have *Kiss My Math*. Thank you so much—we'd recommend this book to anyone!   —Michael Termini, Iowa City, Iowa

We are totally in love with your book! *Kiss My Math* has provided our urban 7th grade students with a fresh, engaging perspective on pre-algebra. Your use of relevant and FUN adolescent examples is something we've never seen before in our combined 21 years of teaching. This book has breathed new math energy into the hearts and minds of our students. Most important is the book's ability to captivate our diverse group of students, regardless of gender, ethnicity, home language, or previous math experience. Thank you!

—Patrice Husak and Ann Hebble,
Murray Junior High School, St. Paul, Minnesota

As a math teacher it's often difficult to find new and exciting ways of explaining concepts—this book provides so many! The explanations in this book have helped some of my students who just "didn't get it" when a concept was explained the traditional way from the textbook. It's amazing; most of my students who read *Kiss My Math* show higher confidence in math—and higher test scores, too.

—Matthew Hartman, algebra teacher,
GW Community School, Springfield, Virginia

Hi Danica, I have your books in my "math library" for students to read in class (I teach 7th and 8th grade math), and they love your books. Last week I caught a student passing what I thought was a note. She showed me what it was after class and it was a clipping about you and *Kiss My Math* from the magazine *CosmoGirl*! Thank you so much for being such a positive role model for young women.

—Ms. Marshall, Southwest Middle School, California

Your books have been the answer to our prayers. My wife and I homeschool our 13-year-old daughter, and she was having a terrible time with math. I bought, downloaded, and sought out every available help that I could find—but everything was so dry that it seemed to make matters worse. Then I read an article about you and it mentioned your books, *Math Doesn't Suck* and *Kiss My Math*, and I immediately brought them home. Our daughter grumbled at first once she heard the word "math," but she started reading and then kept reading and then read some more. She started to see how math was a part of everyday life, and even more miraculously, you've really helped her overcome her *fear* of math. It's been a "math consciousness" breakthrough!

Girls truly speak a different language and you (being a girl yourself) are fluent, which truly allows the math veil to be lifted from their eyes. I can honestly say that to everyone I meet, no matter if their child attends public school or is homeschooled, I sing the praises of your book. You actually *teach* with your books and I cannot thank you enough for that.

—George Foster, Indianapolis, Indiana

Also by Danica McKellar

*Math Doesn't Suck: How to Survive Middle School Math
Without Losing Your Mind or Breaking a Nail*

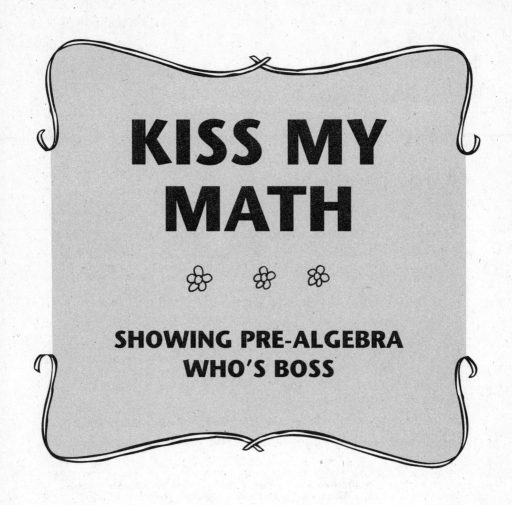

KISS MY MATH

❀ ❁ ❀

SHOWING PRE-ALGEBRA WHO'S BOSS

Danica McKellar

A PLUME BOOK

PLUME
Published by the Penguin Group
Penguin Group (USA) Inc., 375 Hudson Street, New York, New York 10014, U.S.A. • Penguin
Group (Canada), 90 Eglinton Avenue East, Suite 700, Toronto, Ontario, Canada, M4P 2Y3
(a division of Pearson Penguin Canada Inc.) • Penguin Books Ltd., 80 Strand, London
WC2R 0RL, England • Penguin Ireland, 25 St. Stephen's Green, Dublin 2, Ireland (a
division of Penguin Books Ltd.) • Penguin Group (Australia), 250 Camberwell Road,
Camberwell, Victoria 3124, Australia (a division of Pearson Australia Group Pty. Ltd.) •
Penguin Books India Pvt. Ltd., 11 Community Centre, Panchsheel Park, New Delhi –
110 017, India • Penguin Group (NZ), 67 Apollo Drive, Rosedale, North Shore 0632,
New Zealand (a division of Pearson New Zealand Ltd.) • Penguin Books (South Africa)
(Pty.) Ltd., 24 Sturdee Avenue, Rosebank, Johannesburg 2196, South Africa

Penguin Books Ltd., Registered Offices: 80 Strand, London WC2R 0RL, England

Published by Plume, a member of Penguin Group (USA) Inc. Previously published in a
Hudson Street Press edition.

First Printing, July 2009
10 9 8 7 6 5 4 3 2 1

 REGISTERED TRADEMARK—MARCA REGISTRADA

The Library of Congress has catalogued the Hudson Street Press edition as follows:
McKellar, Danica.
    Kiss my math : showing pre-algebra who's boss / Danica McKellar.
        p. cm.
    Includes index.
    ISBN 978-1-59463-049-1 (hc.)
    ISBN 978-0-452-29540-7 (pbk.)   1. Mathematics—Study and teaching (Secondary)—
United States.   2. Girls—Education—United States.   3. Girls—Psychology—United
States.   I. Title.
    QA13.M3185 2008
    512—dc22   2008016124

Printed in the United States of America
Set in ITC Stone Informal
Original hardcover design by Sabrina Bowers

To all the girls who used *Math Doesn't Suck* to kick some serious butt in their math classes and are ready for more— I love your emails, and I'm *so* proud of you!

Acknowledgments

Thank you to my parents, Mahaila and Chris, for always encouraging and believing in me! Thank you to my sister Crystal for just being my best friend in the whole world. And thank you to the rest of my wonderful and supportive family including (but not limited to!) Chris Jr., Connor, Grammy, Opa, Lorna (and kids!), Jimmy, and Molly. Your support means so much and makes everything come full circle for me.

Thank you to the amazing team at Hudson Street Press for making it happen—Clare Ferraro, Luke Dempsey, and my lovely and talented editor, Danielle Friedman. A big welcome to Meghan Stevenson—this will be fun! And thank you to HSP's great publicity team, including the amazing Liz Keenan and Marie Coolman, for all your help and hard work.

Thank you to *all* the fabulous ladies at the Creative Culture, most especially to my wonderful agent, Laura Nolan, for tirelessly translating and smoothing the bumps in the road; I don't know what I would do without you. Thank you to my amazing new publicist Michelle Bega at Rogers & Cowan—your enthusiasm and commitment are priceless! And thank you to Hope Diamond, Brenda Kelly Grant, and Danielle Dusky for years of friendship and guidance! Thank you to my fantastic lawyer Jeff Bernstein and colleagues, especially Leon Liu, Nigel McNulty, and Marie Connolly, and to everyone at CESD for your continued support—including Cathey Lizzio and Pat Brady. Adam Lewis, thank you for everything, and congrats to you and Lori on your beautiful baby girl! Olivia will love math, I promise.

Thank you to Jamie B. for the waltzing breaks. Thank you to my friends for being so patient when I disappeared for several months . . . and thank you for taking me back! Gocha & Shorena, David & Kim, Dan L., Meeghan, and more—you know who you are! Thank you to my goddaughter Tori for keeping me in touch with the life of a 13-year-old, and to Kimberly Stern for more than 20 years of unconditional, unwavering support and friendship. You mean so much to me.

Thank you to my wonderful and resourceful research assistants: my right-hand gal Nicole Cherie Jones, and also Aliza Feldman, Paul A. Jackson, Kristen Ostby, Brittany Pogue-Mohammed, and a big thank you to Anne Lowney for also contributing to the quizzes. You guys rock!

Thank you to Barbara Jacobson—it's been wonderful to reunite. Thank you for putting me in touch with Kay Carlson (my eighth-grade math teacher!) and Karen Salerno—and thank you all for your incredibly helpful feedback. Thank you to Ben Weiss and everyone who helped proofread this book for your great catches and insight, including Anne Clarke, Kari Gulbrandsen, and also Kim, Dina, Ryan, Allen, Yvette, and Andrew, and especially my mom and dad, Brandy Winn, Todd Rowland, Damon Williams, Kayo Goto, and my amazing brother-in-law Mike Scafati.

Thank you to the lovely "testimonial" ladies and all of the teen contributors for sharing your stories. Thank you to all the students across the country who filled out surveys and polls, and to those who gathered them, especially Shirley Stoll and her teachers in Granite City: JoAnn Aleman, Mary Perdue-Tapp, Lisa Miller, Christine Douglas, and Janice Jenek. I also want to thank teachers Bruce Rhodewalt, Denise Hays, David Getz, David DeSpain, JoAnna Kai Cobb, Matthew Hartman, Susan Smith, Rachel Bosworth, Dave Fehringer, Karen Boyk, Marc Ybarsabal, Mary Ann Wiedl, and more! Thank you for your help and insight into math education today.

Also thank you to Eve Newhart, Doc Ogden, and especially Marshall Campbell at Citizens First, and Dan Degrow, Joanne Hopper, Kris Murphy, Sarah Biondo, Jim Licht, and the staff at St. Clair RESA for being incredible supporters of my books—and of math education in general! Thank you to all of my math teachers/collaborators throughout the years, including Lincoln Chayes, Brandy Winn, Barbara Jacobson, and the late Mike Metzger, and thank you to all math teachers everywhere—you are a valuable resource for our young people and are not often fully appreciated for what you do.

And finally, thank you to my loving sweetheart Michael, who keeps thinking of great titles for my books . . .

What's Inside?

PART 4: ALL ABOUT EXPONENTS

PART 5: INTRO TO FUNCTIONS AND GRAPHING LINES

Kiss My What?

"**M**ath? Are you kidding me?"

In high school, a teacher once suggested that I be a math major in college. I thought, "Me? You've got to be joking!" I mean, in junior high, I used to come home and cry because I was so afraid of my math homework. Seriously, I was *terrified* of math.

Things had gotten better for me since then, but still—college math? That sounded really hard; I didn't think I could hack it. Besides, *who* studies math in college other than people who want to be math teachers, right?

Boy, was I *wrong*. Just ask my friend Nina.

In college, Nina wanted to be a doctor more than anything in the world. She'd always wanted to deliver babies! She was smart, funny, and totally capable of doing whatever she set her mind to—until she found out that calculus was a required course. The idea of taking calculus scared her so much that she dropped out of the pre-med program and gave up her dream!

And Nina isn't the only one. Believe it or not, lots of people change their majors and abandon their dreams just to avoid a couple of math classes in college.

So, what does "Kiss My Math" mean?

It means: "Um, excuse me, I'm going to do whatever I want with my life, and I'm sure as heck not going to let a little *math* get in my way."

Who knows what you'll do? Armed with math, you might become a cutting-edge scientist and develop your own line of all-natural makeup or therapeutic high heels. You might discover a cure for cancer or travel into space. You might create some cool engineering trick that destroys trash or creates super clean energy and saves the planet!

Something else, perhaps? Doctor, lawyer, clothing designer, architect? Maybe you'll work for a big magazine or your favorite fashion line, or maybe you'll start your own business.

Believe it or not, all of these fabulous careers use—that's right—*math*.

Check out Stephanie Perry, Jane Davis, and Maria Quiban's real-life testimonials on pp. 37, 71, and 128 to see how studying mathematics gave these ladies a leg up on their competition in the worlds of television, fashion, and magazines. Betcha never knew math could give you power and freedom in *those* areas.

And if anyone tells you it's impossible to be fabulous and smart and make a ton of money using math, well, they can just get in line behind you—and kiss *your* math.

FAQs:
How to Use This Book

What Kinds of Math Will This Book Teach Me?

The chapters of this book are filled with things like breath mints, pandas, popularity, gift wrapping, and spas. By the time you finish reading them, however, you'll be a whiz at tons of pre-algebra topics, including integers, negative numbers, absolute value, inequalities, the distributive property, working with variables, word problems, exponents, functions, graphing, and tons of ways to solve for x. Yep! In fact, these are the topics that tend to be the most confusing, and if you don't understand them now, they can cause tons of trouble later in algebra. That's right—they don't just go away. So let's clear them up now, shall we?

And just to make sure you're *never* confused, every single problem has an answer at the back of this book, as well as a fully worked-out solution on the "solutions" page of kissmymath.com so you can see *exactly* how to do them in case you get a different answer. Sort of makes you feel all warm and fuzzy inside, just knowing that, doesn't it? (Don't answer that.)

What's the Difference Between This Book and Your First Book, *Math Doesn't Suck*?

This is the *next step* in pre-algebra. In *Math Doesn't Suck*, I taught you prime numbers, factors, multiples, fractions, decimals, percents, ratios, rates, proportions, and unit conversions, and introduced you to the idea of variables and solving for *x*—all stuff that prepares you for what is taught in this book. This is the next step up, as you move closer to algebra. You've graduated to the next level, ladies!

Just like in my first book, I've done some of the math in my own handwriting, because I want you to feel like I'm sitting right next to you, helping you to not be confused anymore. I mean, who likes being confused?

What Should I Already Know in Order to Understand This Book?

To get the most out of this book, you'll want to have a good understanding of the topics listed above, like factors, fractions, and decimals. But, I mean, what are the chances you're a total expert on all that stuff? Everyone forgets things!

To make sure that you never feel lost, throughout the book, I've included footnotes that say stuff like, "To review such and such, see p. –– in *Math Doesn't Suck*," so you can quickly flip to it. If you don't own *MDS*, that's fine, too; there are tons of other places to review those topics (like online—just do a search for your topic). This way, though, you're totally covered!

Do I Need to Read the Book from Beginning to End?

Nope! There are a few different ways to use this book:

- You can skip directly to the chapters that will help you with tonight's homework assignment or tomorrow's test.

- You can skip to the math concepts that have always been problem areas to clear them up for good!

- Or you can, in fact, read this book from beginning to end and refer back to each chapter's TAKEAWAY TIPS for quick refreshers as you need them for assignments.

Does "Kiss My Math" Mean, Um, What I Think It Does?

Well, I guess you didn't read the section four pages earlier, now, did you?

What's in This Book Besides Math?

In addition to the math I teach, look out for these fun extras, and more!

- Personality Quizzes: Are You a Stress Case? Do You Pick *Truly* Supportive Friends? Find out now on p. 77 and p. 231!

- Quotes from real teens and famous celebrities.

- What Guys *Really* Think . . . About Smart Girls and other polls taken by students like you. See what everyone's saying!

- Real-life testimonials from gals who overcame their struggles in math and are now fabulously successful women! We've got everything from a fashion buyer to a TV weather anchor. And yes, they all use math in their jobs.

Can This Book Help Me Improve My Test Scores?

Yes! In addition to clearing up any math confusion you might have, I've included a Math Test Survival Guide! at the back of the book. Taking tests is a skill unto itself, and over my many years of math classes in high school and college, I gathered tons of tricks to make the whole thing smoother. Say good-bye to test anxiety—be sure to check it out on p. 305!

Alright, ladies—let's get started!

Chapter 1

Breath Mint, Anyone?

Adding and Subtracting Integers
(Including Negative Numbers!)

Integers

The *integers* consist of the counting numbers, their negative counterparts, and zero. In other words: {. . . −3, −2, −1, 0, 1, 2, 3 . . .}.

. . . I'm already bored.

I'm sorry, but "integer" has got to be the most boring, sterile word I've ever heard. Doesn't it remind you of, I don't know, some type of medical instrument used in a hospital? Some doctor says to her intern: "Quick, pass me the *integer*. We have to operate." Actually, that's already sounding more exciting.

Well, I propose something different. We're not going to talk about these so-called "integers," or I might fall asleep while typing.* Instead, we're going to talk about MINT-egers. Yes, mint-egers, which rhymes with <u>integers</u>, in case you didn't notice. (Sometimes I'll call them "mints" for short.)

. . . .

* Not a pretty sight. Ever woken up with the impression of a keyboard on your face? Okay, fine, me neither, but I'd like to keep it that way.

MINT-EGERS!

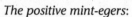

Ah, much better. First let's talk about the positive mint-egers: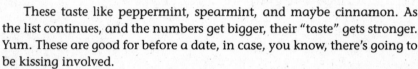

The positive mint-egers:
$$1, 2, 3, 4, 5, 6, 7, 8, 9, 10, 11 \ldots$$

These taste like peppermint, spearmint, and maybe cinnamon. As the list continues, and the numbers get bigger, their "taste" gets stronger. Yum. These are good for before a date, in case, you know, there's going to be kissing involved.

Now, for the dark side of the mint-egers:

The negative mint-egers:
$$-1, -2, -3, -4, -5, -6, -7, -8, -9, -10, -11 \ldots$$

These negative mint-egers taste like those Harry Potter jelly beans with flavors like "vomit," "dirt," and even "booger."* As the list continues, they get stronger and stronger . . . in a bad way. Getting -1 or -2 isn't so bad, but steer clear of -12, yikes!

Even 0 gets to be a mint-eger. He's also known as the "<u>lint</u>-eger," because of his completely bland taste. (Okay, so I've never actually tasted lint, but I bet it wouldn't have much going on in the flavor department.)

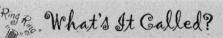

"Ring Ring" What's It Called?

The Integers

Here are the *integers*: $\{\ldots -3, -2, -1, 0, 1, 2, 3 \ldots\}$. Those " . . . " mean that the integers keep going in both directions, forever. Here's what they look like on a number line:

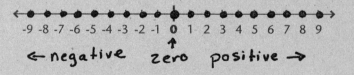

$-9\ -8\ -7\ -6\ -5\ -4\ -3\ -2\ -1\ \mathbf{0}\ 1\ 2\ 3\ 4\ 5\ 6\ 7\ 8\ 9$

← negative   zero   positive →

. . . .

* These jelly beans were clearly designed for little boys, not us.

Notice that *integers* don't include things in between those little marks. So numbers like $1\frac{1}{2}$ and -4.5 are *not* integers. I'd like to think this is because you can't easily break a mint into smaller pieces. Ever tried that? It just crumbles. Very messy.

Back to our mints: The mint-egers get bigger and bigger as you go to the right; their *value* gets bigger. And as you go to the left, their *value* gets less and less. I know this last concept is strange, because the numbers on the left *look like* they're getting bigger, but their *value* is decreasing. A -24 mint has less value than a -1 mint. I mean, wouldn't you rather have a -1 mint than a really yucky -24 mint?

You can also think of "less value" this way: If you saw two houses for sale, and one was in sort of bad shape (a few broken windows, peeling paint) and the second house was *really* falling apart (roof caved in, flooring torn up), which one would be worth more? Worth less? The one that was *more* of a mess would have *less* value, right? For this same type of reason, -24 has a *lesser value* than -1.

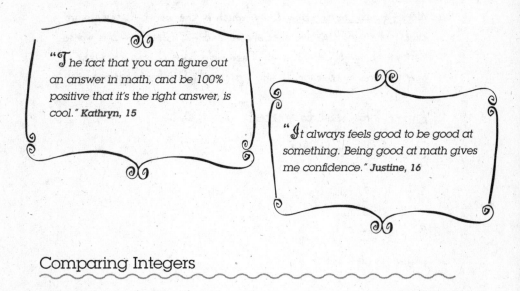

"The fact that you can figure out an answer in math, and be 100% positive that it's the right answer, is cool." **Kathryn, 15**

"It always feels good to be good at something. Being good at math gives me confidence." **Justine, 16**

Comparing Integers

You already know things like:

$$4 < 5$$

The 5 mint has more *value* than the 4 mint. Its value is bigger; in other words, it *tastes* better, and that's why the alligator mouth wants to eat it. Alligators like mints, too; didn't you know that?* However, even alligators know that −5 tastes *worse* than −4:

$$-5 < -4$$

Look at that for a moment until it makes sense. The −5 has a lesser value than the −4. This means that the −4 has a greater value than the −5. Do you see why? Look at the number line again, and see how something to the *right* will always have a greater value than something to the *left*, whether the numbers are positive or negative. Make sense?

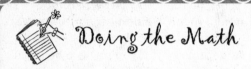 Doing the Math

Arrange these integers in order from *least* to *greatest*. (Think of mint-egers.) I'll do the first one for you!

1. 0, −1, −2, 1

<u>Working out the solution</u>: Okay, which is the worst-tasting mint here? The only negative ones are −1 and −2. Clearly, −2 is worse tasting than −1. So −2 < −1. Then we'd have 0, and then the only positive mint-eger, 1, which is clearly the best-tasting one.

<u>Answer</u>: From least to greatest: −2, −1, 0, 1.

2. −5, 3, 0, −12

3. −4, −7, −10, 6

4. 7, −8, 2, −1

(Answers on p. 318)

- - - - - - - - - -

* For both the "is greater than" symbol > and the "is less than" symbol <, the *open* part is like the alligator mouth, opening to the *bigger* number. That's because the alligator is, um, hungry. We're still doing math. Really.

Combining Integers*

The really convenient thing about these mint-egers is that if you get a negative one in your mouth, like -6, you can immediately *combine* it with a 6 (the positive mint-eger of the same strength), and they neutralize each other! You end up with 0, Mr. Lint-eger himself.

$$-6 + 6 = 0$$

What are some other pairs of mints you could combine in your mouth? Let's say you get that gross -6 in your mouth, and you combine it with a really strong positive mint-eger like 8, then you'd end up with *more* than just a neutral taste, right? You'd actually have a small, nice minty taste left in your mouth.

How *much* of a "small" nice minty taste is left? Well, since 8 is **2** more than 6, if you combine a gross -6 mint with a yummy 8 mint, it's the same as if you'd just eaten a yummy **2** mint to begin with, and nothing else. See?

$$-6 + 8 = 2$$

On the other hand, if you had the negative -6 in your mouth and only had a remedy of a positive 4, then sure, you'd still eat the 4, but you're not going to be totally happy. I mean, the positive 4 *tried* to help fight against the bad taste of the negative -6, but the 4 just wasn't quite strong enough to finish the job, you know what I mean? There was still a bad -2 taste left over. Do you see how this is the same as if you'd started out by eating a gross -2 mint-eger and nothing else?

$$-6 + 4 = -2$$

We also could solve this using the number line: Just think of it as walking back and forth across the number line.

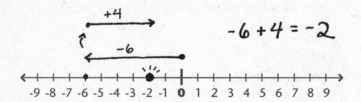

Starting at Mr. Lint-eger, 0 (also called the "origin"), "-6" means we walk in the negative direction 6 steps, and we end up at -6. Then we're

. . . .

* *Combine* is just another way of saying "add," but it's easier to think of it as *combining* now that we're dealing with negative numbers. Trust me on this one!

supposed to add a positive 4, so we walk in the positive direction 4 steps, and we end up at −2.

Try drawing your own version of −6 + 8 = 2 on the number line using the arrows. Remember to start at zero.

Also, we could have done either of these problems starting with a *positive* mint and then combining it with a negative one: 8 + (−6) = 2.

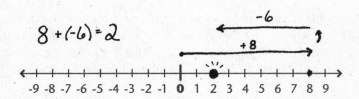

$$8 + (-6) = 2$$

How about something like this: 3 − 5 = ? What could that mean? First, let's do a much more familiar problem: 5 − 3 = ? Well that's easy; the answer is 2. But how did we actually do that? Looking at the number line, it looks like we combined 5 with −3. So, that's: 5 + (−3).

$$5 + (-3) = 2$$

In the same way, we can do the problem 3 − 5 by rewriting it as 3 + (−5).

$$3 + (-5) = -2$$

So we've shown that 3 − 5 = −2.

QUICK NOTE   These all mean the same thing:

$$3 \qquad \text{three} \qquad +3 \qquad \text{positive } 3$$

So $+2 + 4 = +6$ is the same as $2 + 4 = 6$.

I don't usually use extra plus signs, but some textbooks do. They'll say $+3 + 5 = +8$ instead of $3 + 5 = 8$. Just so you know, they mean the <u>same thing</u>.

Rewriting Subtraction → "Adding Negatives"

From now on, whenever we see two positive numbers separated by subtraction, like $3 - 5$ or $9 - 15$, it's best to rewrite the subtraction as "adding a negative." So $9 - 15 = 9 + (-15)$. That way, you're in a position to combine the mint-egers like we've been doing.

Notice that using parentheses around the -15 makes everything easier to read. Otherwise you'd have $9 + -15$, and that looks sort of confusing. I mean, what if you were sloppy and when you wrote "$+-$" it looked more like a plus sign with an extra long line, like this?

$$9 +\!\!- 15$$

So always use parentheses to separate negative numbers from other symbols!

Bigger Numbers

What if we had $28 - 37 = ?$

Before we do anything, let's get rid of the subtraction and rewrite it as "adding a negative": $28 - 37 = 28 + (-37)$. On the number line, we'd go to positive 28, and from there, we'd walk in the negative direction 37 steps. We can tell we'll end up with a negative answer, but *where* does the answer fall on the negative part of the number line? Easy! We simply reverse the order of the numbers and do the "regular" subtraction problem in the margin: $37 - 28 = 9$, and because we know our answer has to be negative, we know our answer is **-9**.

On the other hand, if we instead have $(-28) + (-37) = ?$, we can see that, because both mint-egers are negative, we start off going to -28 and then walk 37 *more* steps in the negative direction. In other words, we can just add $28 + 37 = 65$, and, remembering that we are moving in the negative direction, we know our answer has to be negative: **−65**. As long as you keep your brain on straight, you can just use regular addition and subtraction to do these problems!

QUICK NOTE   If you want to combine more than two integers together, like this: −9 + 6 + (−8) = ? . . . start by combining just the first two numbers. So, pretend nothing else in the world exists, and combine −9 + 6 = −3. Then rewrite the problem and finish it: −3 + (−8) = −11.

Step By Step

Combining integers separated by + and − signs:

Step 1. Rewrite the problem, if necessary, so that all subtraction is changed into "adding negatives." Be sure to surround the negative numbers with parentheses.

Step 2. Going from left to right, begin by focusing on just the first two numbers.

Step 3. If you are combining two negative numbers, *add them* as if they were positive, and then stick the negative sign on afterwards.

Step 4. If you are combining a positive number with a negative number, *subtract them* as if they were positive. Then, remember which "mint" was stronger,* and the sign of the stronger mint will be the sign of the answer.

Step 5. If there are more numbers, keep going!

. . . .

* By the way, the formal way to define the "strength" of a number is something called "absolute value," the number's distance from zero. We'll learn about that in Chapter 4!

QUICK NOTE   In Step 1, you can surround all negative numbers with parentheses, even the first one (which doesn't actually need the separation), if you want. So, −5 − 5 − 5 can be written as either −5 + (−5) + (−5) or (−5) + (−5) + (−5). You can do whichever you prefer, and you'll see it done both ways!

 Doing the Math

Combine these integers, changing subtraction to "adding a negative" when you need to. I'll do the first one for you.

1. 9 − 15 + 7 = ?

<u>Working out the solution:</u>

Step 1. Let's rewrite it so there's no subtraction, just "adding negatives": 9 + (−15) + 7.

Step 2. Then, we'll begin by focusing only on the 9 + (−15) part of the problem and ignore the rest.

Steps 3, 4, and 5. Let's see. We have one positive and one negative number. Because the −15 is stronger tasting than the 9, we know we'll get a negative answer. First, we subtract: 15 − 9 = 6. And now we can stick on the negative sign: −6. So our full problem ends up being −6 + 7 = 1.

<u>Answer:</u> 9 − 15 + 7 = 1

2. 3 − 5 + 4 = ?

3. −3 − 5 + 4 = ?

4. −3 − 5 − 4 = ? (Remember: After you rewrite the subtraction as "adding negatives," do just two numbers at a time, and you'll be fine!)

(Answers on p. 318)

More Ways to Think About Negative Numbers

Let's face it: The whole idea of negative numbers can be really confusing. I mean, how can you have a *negative* number? You can't have −3 apples, after all. But you *can* be 3 feet below sea level or on the P3 level of a big shopping mall. In Europe, they often have a ground level G, and *then* the first floor, second floor, and so on.

When you think about it, an elevator pad is like the number line but in a vertical position. Just pretend that G, ground level, is really zero, P2 is the same as −2, P1 is the same as −1, and so on.

Another common real-world use of negative numbers is the idea of borrowing money, in other words, *debt*. Let's say that you don't have any money and you want to buy a new hair clip. So, you borrow $2 from your friend and you buy the sparkly clip. At this point, you have −$2, because you *owe* $2, and you started out with *no* money. A week later, when you get $10 for your birthday and you pay her back, how much money are you left with? This is $-2 + 10 = 8$. So you have $8 left.

Personally, I'd rather think of it as mint-egers in my mouth, but that's just me.

Whether you prefer to think of mints or sea level or money or elevators, or just walking back and forth along the number line, the important thing is to find something that gives negative numbers some sort of *meaning* to you. It doesn't matter which one you choose; they're all saying the same thing. The best choice is the one that helps a problem like this make *sense* to you: $9 + (-15)$.

"Math definitely isn't my best subject, but if I really, I mean *really* try to understand things, I can succeed." **Joanna, 14**

"Being smart is something you work toward." **Kelsi, 14**

Subtracting Integers

So far, we've been able to avoid the whole "subtracting" thing by changing subtraction into "adding negatives." Why? Because <u>combining mints in your mouth and walking up and down the number line only works when you have addition between each term.</u>

So, what about a problem like $4 - (-3) = ?$ How can we rewrite this so we're combining terms with addition? Is that even possible? I mean, how can we change the subtraction into "adding a negative" when 3 is *already* negative? Hmm.

Luckily, we don't have to worry about any of that! Believe it or not, subtracting a negative number is one of the *easiest* things to do: The two negative signs actually cancel each other out and become a plus sign. And addition is so much nicer, don't you agree? Here's the shortcut:

Shortcut Alert: Subtracting Negative Numbers

When we are asked to *subtract a negative number* like this, $10 - (-5)$, we can just change the two negatives into a positive plus sign! So: $10 - (-5) = 10 + 5 = 15$. Now that's what I call a shortcut.

When the minus sign and the negative sign are next to each other like that, it's almost like one of them turns on its side and moves over until they're touching. Together they create the shape $+$. It's a very special moment for the two of them.

So, the problem $4 - (-3)$ turns out to be really easy! $4 - (-3) = 4 + 3 = 7$. But why does this trick work? Well, after we get through Chapter 3, we can totally answer that question, no problem.* For now, just be glad that subtracting a negative has such a nice shortcut. Yes, this is a good time to practice that warm, grateful feeling that can make life so much happier.

. . . .

* But what does it *mean*, philosophically, to subtract a negative number? Imagine you have a bunch of positive and negative mints in your mouth, for an overall good taste of "4." Then *spit out* a yucky "–3" mint. Guess what? Now your mouth tastes like 7, right? Well, that's 4 – (–3) = 7. See kissmymath.com/extras for more!

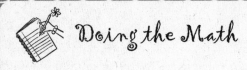 Doing the Math

Use the above shortcut to evaluate these expressions. If you need a refresher, check out the step-by-step method on p. 8 for combining integers. You should apply our shortcut before you even get to Step 1. I'll do the first one for you.

1. $-3 - 4 - (-9) = ?$

<u>Working out the solution</u>: Okay, first we'll go through and use the shortcut to look for double negatives (two negatives in a row) and let them have their magic moment of turning into a positive plus sign. As we can see, that happens right before the 9, so we can rewrite our problem to be $-3 - 4 + 9$. Now we can change the minus sign in front of the 4 into "adding a negative": $-3 + (-4) + 9$. Okay, NOW we're ready to start combining terms. Left to right, we get $-3 + (-4) = -7$. And now our problem looks like this: $-7 + 9 = 2$. Done!

<u>Answer</u>: $-3 - 4 - (-9) = 2$

2. $2 - 4 - (-8) = ?$

3. $-3 - (-7) = ?$

4. $1 - (-2) - 1 = ?$

5. $-1 - 1 - (-1) - (-1) = ?$ *(Hint: Do this methodically; don't skip steps!)*

(Answers on p. 318)

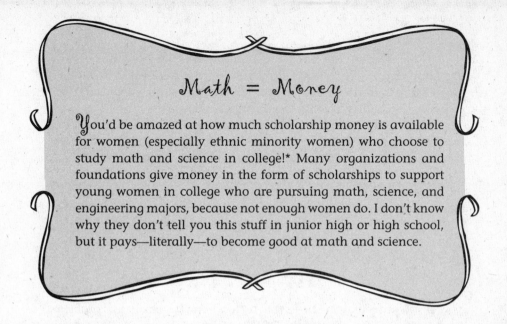

Math = Money

*Y*ou'd be amazed at how much scholarship money is available for women (especially ethnic minority women) who choose to study math and science in college!* Many organizations and foundations give money in the form of scholarships to support young women in college who are pursuing math, science, and engineering majors, because not enough women do. I don't know why they don't tell you this stuff in junior high or high school, but it pays—literally—to become good at math and science.

Beyond Integers: Negative Fractions and Decimals

Has there ever been a guy you really like, but he never even looks in your direction and you're pretty sure he doesn't even know you're alive? Now imagine that one day you're talking with your friends in the hall, and you say something particularly clever and funny. All your friends laugh at your witty joke, and guess who was walking by and heard you, too? Yes, it's *him*! He turns around, amused and impressed by you, your eyes lock (slow motion, music swells), and he smiles right at you and winks! *Bam!* Your heart is soaring, and your cheeks are flushed! If he never noticed you before, he sure has now. What a feeling.

As it turns out, there are lots of values on the number line that we might not have realized existed before. We've been walking right past them in the hall. But now we *will* notice them. And just think about how good they'll feel when we do! Kinda makes you want to notice them even more, doesn't it? Oh, be nice.

For instance, have you ever noticed where $-4\frac{1}{3}$ hangs out? Picture the number line, and *notice* that it would be between -5 and -4, closer to the -4. And how about -1.5? Where does it live?

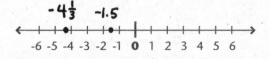

* For more details, check out websites like collegescholarships.org/women.htm.

See, the difference between this example and the integers is that on *this* number line, we're paying attention to the numbers that happen *between* those little tic marks. The values were there all along between the integers; we just didn't notice them before.

The good news is that they obey all the same "combining" rules as integers! (And yes, they obey the same rules as our mint-egers, too.) So you can use the same exact step-by-step methods for these numbers as you did on p. 12.

For whatever reason, problems with fractions and decimals can tend to feel a little muddy. Sometimes just *looking* at them can make you feel panicky. But remember: Once you rewrite them so there's no more subtraction and you're left with "combining negatives," it's like mint-egers in your mouth again!

$$-8.5 - 9.3 - (-4.35)$$

Now, if you immediately know what to do when you look at this problem, great—you can skip down to the DOING THE MATH on p. 16. If the above problem makes you feel a little uneasy (and that's totally normal), then read on!

Got a Scary Problem? Try the _Easier_ Version!

Does this look a little scary?

$$-8.5 - 9.3 - (-4.35)$$

Here's my advice for problems like this. First, write down the easier version of the same problem on a scratch sheet, the only difference being that you use *easier numbers* (plain integers; no fractions or decimals). So, on a totally separate piece of paper (not on your homework), you'd write this:

$$-8 - 9 - (-4)$$

This should make it easier to "see" the method you should use to start, step by step.

We see a double negative in front of the 4, so we can turn it into a plus sign, right? And then we can change the subtraction in front of the 9 into "adding a negative," $-8 + (-9) + 4$, which leaves us in a nice familiar situation, like we're combining mints in our mouth on p. 5.

So now, on your homework, do the same thing with the non-integer numbers: First, notice that the double negative in front of 4.35 can become a positive, and change the subtraction in front of the 9.3 into "adding a negative":

$$-8.5 + (-9.3) + 4.35$$

Okay, good progress, we're doing great. Now: Left to right, let's attack the first two terms and pretend the third one doesn't exist: $-8.5 + (-9.3)$.

Did the brain lock up again? Happens to the best of us. If it did, then it's time to go back to the scratch paper and figure out the method for evaluating an *easier version* of this. How about $-8 + (-9)$?

In this expression, you can see that we are combining two negative numbers, so we need to go even farther down the negative side of the number line, right? You start at -8 and go even farther to the left. To see where we end up, just add the two numbers without their negative signs, $8 + 9 = 17$, remembering that the actual answer is negative: $-8 + (-9) = -17$.

Using the *same method* for $-8.5 + (-9.3)$, we know that we can just add the two negative numbers *without* the negative signs: $8.5 + 9.3 = 17.8$. Yet, knowing our answer has to be negative, we'd write $-8.5 + (-9.3) = -17.8$.

So, how does our problem look now? It's become:

$$-8.5 - 9.3 - (-4.35)$$
$$\downarrow \qquad\qquad \downarrow$$
$$-17.8 + 4.35$$

Looking a little better, right? Again, how would this problem look on our "easy paper"? We could consider $-17 + 4$. To combine these mint-egers, we know that we'll end up with a negative taste in our mouth at the end. So we can subtract $17 - 4 = 13$, and then give it the *sign* of the stronger mint, in this case *negative*. So: $-17 + 4 = -13$.

Applying this to our real problem: To combine $-17.8 + 4.35$, we can just subtract $17.8 - 4.35 = 13.45$, and then, because the stronger mint was negative, we know our answer must be negative, so: $-17.8 + 4.35 = \mathbf{-13.45}$. And we're done!

That took so long only because I talked and talked and talked so that there would be no confusion at all. Hopefully, you now feel armed with a technique for handling problems that seem scary at first—by doing an easier version on the side!

Depending on the kinds of problems your teacher assigns, you might want to brush up on your addition and subtraction for decimals and fractions (all that stuff with finding common denominators, etc.). After all, if you're doing a problem like this, $\frac{1}{6} - \frac{14}{15}$, the last thing you need to be worried about is trying to remember how to find the LCM of 6 and 15!* I do a whole big review of fractions in Chapter 8 of *Math Doesn't Suck* and decimals in Chapter 10 of *Math Doesn't Suck*, in case you want to check them out!

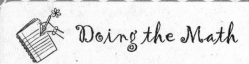 Doing the Math

Evaluate these expressions. Remember: The same rules apply for fractions and decimals as for integers. Refer to the STEP BY STEP on p. 8 if it helps. I'll do the first one for you!

1. $-\frac{1}{2} - (-4) - \frac{5}{2} = ?$

- - - - - - - - - -
* I just can't resist showing you the subtraction:
$$\frac{1}{6} - \frac{14}{15} = \frac{5}{30} - \frac{28}{30} = \frac{5 - 28}{30} = \frac{5 + (-28)}{30} = \frac{-23}{\mathbf{30}}$$

<u>Working out the solution</u>: The first thing to notice is that we have two negatives in a row in front of the 4; let's change them into a single positive sign, using our shortcut: $-\frac{1}{2} + 4 - \frac{5}{2}$. Next, we're going to get rid of the subtraction and change it into "adding a negative." So now our problem looks like this: $-\frac{1}{2} + 4 + \left(-\frac{5}{2}\right)$. Make sure you understand everything so far. Now, only + symbols are between each term, and that's how we like it: We're ready to combine! Let's start with the first two terms: $-\frac{1}{2} + 4$. One is positive and one is negative, so we'll subtract them. The bigger one is positive, so we'll end up with a positive answer: $4 - \frac{1}{2} = 3\frac{1}{2}$. That makes sense, right? Now the problem looks like this: $3\frac{1}{2} + \left(-\frac{5}{2}\right)$. Rewriting $3\frac{1}{2}$ as an improper fraction, $3\frac{1}{2} = \frac{7}{2}$, the full problem is: $\frac{7}{2} + \left(-\frac{5}{2}\right)$. One is positive and one is negative, so we can just subtract them. The positive one is bigger, so we'll end up with a positive answer: $\frac{7}{2} - \frac{5}{2} = \frac{7-5}{2} = \frac{2}{2} = 1$. And that's it. We combined all the terms, so we're done!

<u>Answer</u>: $-\frac{1}{2} - (-4) - \frac{5}{2} = 1$

2. $-\frac{1}{2} - (-3) - \frac{1}{2} = ?$

3. $43.3 - 56.9 = ?$

4. $43.3 - 56.9 + 2.6 = ?$

5. $\frac{5}{2} - \frac{7}{2} - (-0.5) + 0.5 = ?$

(Answers on p. 318)

Takeaway Tips

 To understand what negative numbers "mean," use whichever example makes the most sense to you. You can envision a number line, or think of an elevator, sea level, money, or breath mints! Go back to it whenever you feel lost or confused.

 Before doing any combining of integers, be sure to do these two things:

1. When you see two negative signs in a row, use the shortcut to turn them into a positive + sign.

2. Then, if there's any subtraction left, turn the minus sign into "adding a negative."

Once all the terms have been written with + signs between them, you are free to think of the problem as *combining mints*, and then combine them left to right.

Negative fractions and decimals work the same way integers do. Just be extra careful with these. And remember: You can always work the problem with "easy numbers" on a scratch sheet to help you keep track of what methods you want to use.

Ugly Yourself Up?

Have you ever dumbed yourself down? Pretended to be *less* than you really are just to fit in? When you think about it, doing this doesn't make any sense. I mean, would you ever "ugly yourself up"? Imagine getting ready for school and putting brown eyeshadow under your eyes to look like dark circles, and not brushing your teeth or your hair. Yeah—not making a lot of sense.

Of *course* we want to be the best we can be—inside and out. We're never going to be perfect, whatever that means, but we might as well give it our all, right?

To have clear skin and a normal weight, we know we should eat healthy food, drink lots of water, get eight to nine hours of sleep (sometimes I need 10; the more I sleep, the better my skin is!), and exercise regularly. And, oh yeah, not deliberately ugly ourselves up with eyeshadow under our eyes.

To stay smart and savvy, we know we should pay attention in class, start big projects far ahead of the deadline, study our notes, ask for help when we need it, and *certainly* not deliberately dumb ourselves down.

Again, no one's perfect, but when we strive to live up to our full potential, inside and out, it feels great and we can adore ourselves for it!

Chapter 2

The Popular Crowd
The Associative and Commutative Properties

$\mathcal{U}$gh. I remember when I had to learn these number properties. I totally had an attitude about it. I was like, "Who needs the commutative property?" I mean, 4 + 5 is the same as 5 + 4? Big woopdie-doo. Why do they have to give it such a long name, which I now have to *memorize*, thank you very much.

And what's up with the associative property? No kidding, (4 + 5) + 6 is the same as 4 + (5 + 6)? Well my goodness, somebody call the newspapers. As far as I could tell, this was just my math book's excuse to make me learn more math vocabulary—a complete waste of my time.

So let me just say that I know where you're coming from. But now I'll let you in on a little secret: Believe it or not, these rules were designed to help us get around the whole PEMDAS* thing and make life easier later on so you can have choices about the *order* you simplify expressions in.

Order of Operations Review

Because we're about to go messing with moving around parentheses and things, this is a good time to review the **order of operations**—and the dining habits of pandas.

. . . .

* PEMDAS is that Order of Operations rule, which we are about to review!

What's It Called?

Order of Operations

The *Order of Operations*, often called PEMDAS, is the order in which we must always simplify a math expression:

Parentheses

Exponents

Multiplication & **D**ivision—whichever one comes first
(Notice that we could have said "Division and Multiplication"; they have the same priority.)

Addition & **S**ubtraction—whichever one comes first
(Notice that we could have said "Subtraction and Addition"; they have the same priority.)

This PEMDAS thing is fine, but personally, I prefer to think about pandas.

Ah, pandas. Aren't they cute? And they have really big appetites. I've heard that pandas like to eat dumplings with mustard, and then for dessert, they have apples with spice. Yum!*

Pandas

Eat

Mustard on

Dumplings and

Apples with

Spice

I recommend saying this out loud a few times until you've learned it. You'll see that it's got a rhythm to it: Pandas eat mustard on dumplings and apples with spice! Notice that the panda eats two different courses: dinner and then dessert. Multiplication and division happen together at dinner, so they have the same priority; we do whichever one comes *first*,

. . . .

* The truth? Pandas eat mostly bamboo. But they do come from China, and Chinese dumplings are yummy, especially with hot mustard. And pandas eat apples in the zoo, so I'm not WAY off. Besides, this'll help you remember how to use the Order of Operations correctly, and that's sort of more to the point, isn't it?

left to right. (So I could have said "dumplings with mustard" instead of "mustard on dumplings.")

The same is true for addition and subtraction (apples and spice): They happen together during dessert, so they have the same priority as each other, and we do whichever one (addition or subtraction) comes *first*, left to right.

PEMDAS might as well have been called PEDMAS or even PEDMSA—see what I mean? Let's use the Panda rule to correctly simplify the expression:

$$36 \div (2 + 1) - 2 \times 4 - 4 + 1 = ?$$

(Watch out: We mustn't be tempted to subtract $4 - 4$ right off the bat. We have to obey the order of operations, and the **S**pice isn't until dessert!)

What's first? The **P**andas. We first simplify inside the **P**arentheses, and $2 + 1 = 3$, so we get:

$$36 \div \mathbf{3} - 2 \times 4 - 4 + 1 = ?$$

The next word is "**E**at," but there are no **E**xponents,* so we move on to the first panda meal: **D**umplings with **M**ustard. It's time to find all the **M**ultiplication and **D**ivision and do them left to right.

Left to right, **D**ivision comes first in this example. So, because $36 \div 3 = 12$, we get:

$$\mathbf{12} - 2 \times 4 - 4 + 1 = ?$$

Next we do the **M**ultiplication; $2 \times 4 = 8$, so we get:

$$12 - \mathbf{8} - 4 + 1 = ?$$

Then we do the **A**ddition and **S**ubtraction, left to right. In this example, subtraction comes first. Because $12 - 8 = 4$, we get:

$$\mathbf{4} - 4 + 1 = ?$$

Continuing to go in order, we now can subtract $4 - 4 = 0$, and we're left with **1**. So the answer is: $36 \div 3 - 2 \times 4 - 4 + 1 = \mathbf{1}$.

Just try doing the operations in some other order, like starting with subtracting the $4 - 4$, and you'll see that you get a totally different, wrong answer. (For more about why this happens, see p. 35. We got the *correct* answer by doing the operations according to the Panda rule.)

Now that we've got the rules straight, let's see how we're allowed to mess with them and still get the right answer!

. . . .

* We'll be learning about exponents in Chapters 15 and 16.

The Associative Properties

Do you have cliques at your school? You know how some girls will only associate with certain other girls? At most schools, this is the case; people hang out in groups. Of course, sometimes these groups change, usually for pretty silly reasons. Sometimes groups change because someone joins the cheerleading squad, someone else starts dating a new guy, or maybe, I don't know, because the weather changed.

Whatever the reason, sometimes girls decide to "associate" with someone new or stop associating with someone they used to like.

Sometimes numbers change their *associations*, too.

For example, let's say that 4 and 2 are best friends today, but they don't care much for 7, because she's always bragging about how tall and thin she is. You might see: $(4 + 2) + 7$.

But maybe next week, 2 and 7 will become best friends because they'll realize they are both prime . . . so I guess so have a lot to talk about, and 4 just wouldn't understand. Now you might see $4 + (2 + 7)$. The associative property of addition says that these two expressions have the same value:

$$(4 + 2) + 7 = 4 + (2 + 7)$$

There's an associative property for multiplication, too, so:

$$(4 \times 2) \times 7 = 4 \times (2 \times 7)$$

After all, both sides equal 56. Try it for yourself, and remember to do whatever's inside the parentheses first.

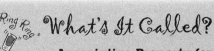

"Ring Ring" What's It Called?

Associative Property for Addition

If a, b, and c stand for any numbers, then:

$$(a + b) + c = a + (b + c)$$

This says that as long as the **only operation is addition**, then you can move around the parentheses, and it doesn't change the *value* of the expression!

For example, $(3 + 2) + 1 = 3 + (2 + 1)$. They both equal 6.

Associative Property for Multiplication

If *a*, *b*, and *c* stand for any numbers, then:

$$(a \times b) \times c = a \times (b \times c)$$

This says that as long as the **only operation is multiplication**, then you can move around the parentheses, and it doesn't change the value of the expression!

For example, $(6 \times 4) \times 2 = 6 \times (4 \times 2)$. They both equal 48, after all!

QUICK NOTE   Remember: The placement of parentheses matters because in the order of operations, we always do what's inside the parentheses first!

Watch Out!

The associative property does NOT work for subtraction or division. For example:

$$(3 - 2) - 1 \neq 3 - (2 - 1)$$

Just try it and you'll see that each side gives a different number! There is no associative property for division, either. For example:

$$(36 \div 6) \div 2 \neq 36 \div (6 \div 2)$$

Again, you'll get two totally different answers for each side. Just remember: There's an associative property of addition and an associative property of multiplication, and that's it!*

The associative property can come in handy when you're multiplying numbers in your head. Check it out . . .

. . . .

* For a sneaky way to use the associative property of *addition* when you're stuck with subtraction, see p. 33. Can you guess how to do it?

Reality Math

Let's say you're in charge of the set decoration for the school play . . . along with the super cute, brilliant new guy in school. You tell him that the budget will allow for up to 48 square feet of material for the curtains. He tells you that you'll need curtains for two small windows, and that each window will be 4.5 feet wide and 5 feet tall (for the play, not for your future house; stay focused!). Now the question is: Will that be less than 48 square feet? Can you guys afford it?

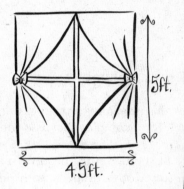

So you think to yourself: *Hmm. For each window, the square footage can be calculated by multiplying 4.5 × 5. Then after I figure that out, I'd need to multiply it by 2 for the total square footage, because there are two windows.* In the meantime, the guy has written down what you were just thinking:

$$(4.5 \times 5) \times 2$$

You probably don't know off the top of your head what (4.5 × 5) is, but you might notice that the associative property of multiplication says it's okay to just move the parentheses to rewrite it!

$$4.5 \times (5 \times 2)$$

You know that (5 × 2) = 10, so now it's easy to multiply 4.5 × 10 in your head and get the answer: 45!

You smile and casually say, "So I guess that would be 45 square feet. Yeah, we can afford that." Cute brilliant guy, trying to hide how impressed he is, simply smiles back . . . and asks to join you at the fabric store.

So, instead of always doing what's inside the parentheses first, we now have *choices*: With these associative properties, we can sometimes move the parentheses without messing up the value of the expression, making some problems much easier! Once we've done that, then we continue with the Panda PEMDAS rule in the normal way. It's like we've learned how to break the rules and not get in trouble . . .

QUICK NOTE   Parentheses ( ) and brackets [ ] can be used interchangeably; they do the same thing!

Negative Numbers

Yes, the associative properties work when negative numbers are involved, too! For example, the associative property of addition means that $-2 + (4 + 5) = (-2 + 4) + 5$. Just try it! Both sides equal 7.

But don't confuse negative numbers with subtraction. It's okay to use the associative property with negative numbers and *addition*, but *not* with subtraction. So, the associative property tells us that $3 + [-7 + 2] = [3 + (-7)] + 2$, because <u>the only operation is addition</u>. It's okay to have negative numbers—as long as they are being <u>added</u>. On the other hand, subtraction is *not* associative, so $3 - [7 + 2]$ is *not* equal to $[3 - 7] + 2$. See the difference?

Remember: For the associative properties to work, make sure the *operations* are either only addition, or only multiplication. (We'll learn how to *multiply* negative numbers in Chapter 3.)

"*I* used to think that smart girls were the nerdy looking girls who did extra homework every night because they thought it was fun and cool. But now I realize that any girl can be a smart girl, no matter what she looks like. One of the most popular girls in school can also be one of the smartest girls in school." **Chelsea, 13**

Doing the Math

Don't start these problems with the Panda Order of Operations rule! Instead, *using the associative properties*, move the parentheses to make these problems easier, and then evaluate them, as usual. I'll do the first one for you.

1. $\left(7 \times \frac{1}{2}\right) \times 2 = ?$

<u>Working out the solution</u>: Sure, we could multiply $7 \times \frac{1}{2}$, but then we get into fraction multiplication. It wouldn't be terrible, but it's much easier, in this case, to move the parentheses, using the associative property of multiplication, and rewrite the same value like this: $7 \times \left(\frac{1}{2} \times 2\right)$. Since $\frac{1}{2} \times 2 = 1$, our answer is $7 \times 1 = 7$.

<u>Answer:</u> $\left(7 \times \frac{1}{2}\right) \times 2 = 7$

2. $-39 + (39 + 58) = ?$

3. $(9 \times 2) \times 5 = ?$

4. $3 \times \left(\frac{1}{3} \times 8\right) = ?$

(Answers on p. 318)

As important as knowing *how* to use the associative properties is knowing *when* you can and can't use them. Try this next set of exercises to see how well you understand this.

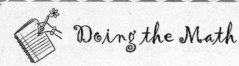

Doing the Math

In each of these problems, the *only* difference between the two expressions is the placement of the parentheses. For each, do the following: **a.** Decide: Does the associative property tell us that the two expressions will be equal? **b.** Evaluate both expressions and see for yourself if they are equal! I'll do the first one for you.

1. $3 + (5 \times 2)$ and $(3 + 5) \times 2$

<u>Working out the solution:</u> **a.** At first, this may look like some sort of associative property, but we have addition and multiplication mixed together, so it's not! So the associative property does *not* tell us if these two expressions will be equal. **b.** Let's evaluate each one. We should get different answers. For the first expression, obeying the Panda rule, we first work inside the parentheses, and we get $3 + (5 \times 2) = 3 + 10 = 13$. For the second expression, again starting inside the parentheses, we get $(3 + 5) \times 2 = 8 \times 2 = 16$.

<u>Answer:</u> **a. no; b. 13 and 16**

2. $7 + (2 + 1)$ and $(7 + 2) + 1$

3. $(6 - 2) + 4$ and $6 - (2 + 4)$

4. $5 \times \left(\frac{1}{2} \times 6\right)$ and $\left(5 \times \frac{1}{2}\right) \times 6$

5. $6 + (2 \times 3)$ and $(6 + 2) \times 3$

6. $(8 \div 2) \div 2$ and $8 \div (2 \div 2)$

(Answers on p. 318)

Danica's Diary

THE POPULAR CROWD

I remember in middle school there was this girl named Jasmine* who used to keep a list of the "popular" girls in her binder. These were the only girls she would associate with. She updated the list every day . . . it was so stressful! Thinking back, there's no good reason why she had any authority to decide who was cool and who wasn't, but I guess we all believed her because, well, everyone else believed her.

One day, Jasmine decided that I was having a very "uncool" day, and she instructed her best friend Veronica, and the other girls on the list, not to associate with me. (If we were numbers, I'd have been the one all by itself, while the others were in parentheses.) But during recess, Veronica boldly declared, "I want to talk to Danica" and marched over to where I was sitting alone on a bench. It was a moment worthy of swelling background music and slow-motion effects. I'll never forget that. Veronica's boldness only lasted one day, but it still meant a lot to me.

The end of the story is this: A few years later, after I had become famous from my part on TV's <u>The Wonder Years</u>, I was going to a different high school, and I ran into Jasmine at a multischool event. Big surprise--she acted like we'd always been the best of friends, giving me her number, etc. She wanted to "associate" with me *then*, didn't she? It's so weird how much that would have meant to me a few years earlier and how little it mattered then. Boy, can a few years change your perspective on people! By the way, I was very polite . . . but I never did call her.

. . . .

* I've changed the names of the girls to protect them . . . I guess I'm just nice that way.

Now that we're experts at the associative properties, we're going to learn about another kind of property. You know all those cool juniors and seniors who brag about driving? Well, someday they'll be complaining about their commute.

The Commutative Properties

Have you ever heard people complain about their commute? Maybe they live in the suburbs but they work in the city, so they have to make an hour-long *commute* every morning to work and then another hour back home in the evening. Yikes! That's a lot of moving back and forth every day!

Well, in the "<u>commute</u>"-ative properties, instead of *cars* moving back and forth, *numbers* move back and forth. For addition, it means that 4 + 5 gives you the same answer as 5 + 4. Or, in math language, we could say that 4 + 5 = 5 + 4.

There's also a commutative property of multiplication, which means things like this: $2 \times 8 = 8 \times 2$. Again, you probably knew that from your times tables. After all, $8 \times 2 = 16$, and $2 \times 8 = 16$, right?

Just like with the associative properties, the commutative properties work just fine for negative numbers, too: −9 + 2 = 2 + (−9). See what I mean? After all, when you're combining mint-egers, it doesn't matter which order you put them in your mouth, you'll end up with the same overall taste!*

Because the commutative properties work for *all* numbers, we can write out the rules using letters to act as nicknames for whatever numbers we might want to use.

. . . .

* See p. 1 if you don't know what the heck a mint-eger is.

"Ring Ring" What's It Called?

The Commutative Property for Addition

For all numbers a and b:

$$a + b = b + a$$

For example, $-13 + 5 = 5 + (-13)$. They both equal -8, after all!

The Commutative Property for Multiplication

For all numbers a and b:

$$a \times b = b \times a$$

For example, 3×4 equals 4×3. They both equal 12, after all!

I used to think these properties were called the "communative" properties or something (that's not even a word, by the way!), instead of the correct term: the *commuTative* properties. But just think of the morning *commu<u>t</u>e* from the suburbs to the city, and you'll remember that first T! Also, if you were going to write a poem about it, *commute* rhymes with *cute*.*

Watch Out!

There's no commutative property for subtraction or division. Just try it!

Is this true: $5 - 4 = 4 - 5$? Nope! After all, $5 - 4 = 1$, but $4 - 5 = -1$. And how about division? Does $8 \div 2 = 2 \div 8$? Nope, again.

After all, $8 \div 2 = \mathbf{4}$, but $2 \div 8 = \frac{2}{8}$, right? Then reduce: $\frac{2}{8} = \mathbf{\frac{1}{4}}$. And last time I checked, 4 does not equal $\mathbf{\frac{1}{4}}$!

• • • • • • • • • •

* I don't expect any of you to actually write me a poem about the commutative property. Do you have any idea how surprised I would be if such a poem showed up in my email? Really, REALLY surprised.

What Do You Really Think?

In an anonymous poll, we asked more than 200 girls ages 13–18 to respond to the following question, and here's what they said. Where do you fall?

Do You Feel Confident Participating in Math Class?

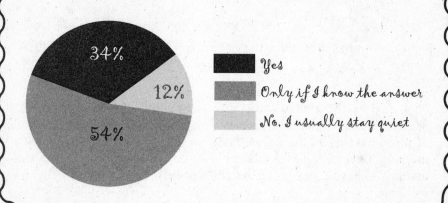

- **Yes** — 34%
- **Only if I know the answer** — 54%
- **No, I usually stay quiet** — 12%

Associative and Commutative Properties—Together!

To review: In the associative properties, the *parentheses* move. Like cliques at school, the numbers change who they're associating with! In the commutative properties, the numbers *themselves* do the moving—like cars going back and forth on their daily commute.

We could summarize how to use the associative and commutative properties like this: If you have *only* addition or *only* multiplication, you can mix up the order and groupings of the numbers however you want and not alter the value of the expression.

Because these properties work for all numbers, they also work with variables, which, after all, are just numbers whose value we don't know yet. We could use both properties to make this look a lot better: $7 + [x + (-7)]$. But instead of using an x, let's use a prettier variable, like 🎴. Why not? It's a free country.

$$7 + [🎴 + (-7)]$$

Using the commutative property of addition, we can switch the order of and −7:

$$7 + [(-7) + \text{❀}]$$

Using the associative property of addition, we can move the brackets, and then we can combine integers and finish:

$$[7 + (-7)] + \text{❀}$$
$$= \text{❀}$$

So if we were using a "real" variable, then our answer would be x, which looks a lot nicer than $7 + [x + (-7)]$ but has the *same value*. Yep, $7 + [x + (-7)] = x$.

The beauty of it is this: We don't have to know the value of x. We're using the properties correctly, so we know that we won't mess up the value of the whole expression—no matter what the value of our variable is!

QUICK NOTE   Sneaky Way to Use the Properties with Subtraction

The only way to use these properties with subtraction is to first write all of the subtraction as "adding negatives" (see p. 7). So, even though we cannot use the properties to rearrange [5 − 3] − 2, we can use the properties to rearrange [5 + (−3)] + (−2), because now, the only operation is addition. Pretty sneaky, huh? (For more on this, see p. 35.)

Doing the Math

Use the commutative and associative properties to make these expressions simpler. (Use flowers instead of variables if you want!) If you're not *allowed* to use the properties (because the problem isn't *all* addition or *all* multiplication), then say "not allowed." I'll do the first one for you.

1. $\frac{1}{5} \times (b \times 10)$

<u>Working out the solution</u>: The problem involves *only* multiplication, so we're allowed to use the properties to move things around. To simplify this, we'll want to get the two numbers together. Using the commutative property, we can switch the order of b and 10 to get $\frac{1}{5} \times (10 \times b)$. Then we'll use the associative property to move the parentheses: $\left(\frac{1}{5} \times 10\right) \times b$, and since $\frac{1}{5} \times 10 = \frac{10}{5} = 2$,* we end up with a final answer of $2 \times b$.

<u>Answer</u>: $\frac{1}{5} \times (b \times 10) = \mathbf{2 \times b}$

2. $(-18.5 + y) + 8.5$

3. $(-18.5 \div y) + 1.5$

4. $7 \times \left(z \times \frac{1}{7}\right)$

(Answers on p. 319)

For more number properties and also the "formal" sets of numbers, see pp. 312–4 in the Appendix. They're pretty easy, so I didn't need to put them in this chapter. They also didn't have anything to do with popularity . . .

Takeaway Tips

To remember the correct order of operations, think of pandas at mealtime: <u>P</u>andas <u>E</u>at <u>M</u>ustard on <u>D</u>umplings and <u>A</u>pples with <u>S</u>pice! The dinnertime operations (<u>M</u>ultiplication and <u>D</u>ivision) have the same priority as each other and should be done left to right. The same is true for the dessert operations (<u>A</u>ddition and <u>S</u>ubtraction).

* To brush up on fraction multiplication, see Chapter 5 in *Math Doesn't Suck!*

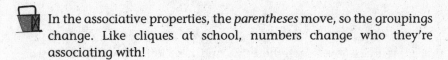 In the associative properties, the *parentheses* move, so the groupings change. Like cliques at school, numbers change who they're associating with!

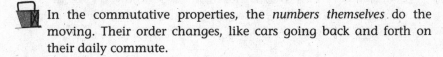 In the commutative properties, the *numbers themselves* do the moving. Their order changes, like cars going back and forth on their daily commute.

Remember: The associative and commutative properties work *only* when an expression is written in terms of only addition or only multiplication.

What's the Deal?
PEMDAS *Mystery* <u>*Revealed*</u>!

This section is totally optional, but it's really intriguing, and it will make you more powerful in math.

Okay, so you know from PEMDAS (see p. 21) that we're supposed to do addition and subtraction together, left to right, and we *never* skip around, because that could be dangerous, even in a problem as easy as $7 - 2 + 1 = \mathbf{6}$. We *have* to subtract the $7 - 2$ first, because it comes first, left to right. If we did the $2 + 1$ addition first, we'd end up with $7 - \mathbf{3} = 4$, which is wrong, because we know the real answer is 6. Our answer of 4 would be too small; *we somehow subtracted more than we should have.* What's up with that?

Well, you may now be able to appreciate *why* this sort of thing happens. Let's rewrite this problem, changing the subtraction into "adding a negative": $7 + (-2) + 1$. Looking at it, you can see that the *only* negative contribution comes from the 2. Everything else is positive minty-ness! Now that it's written like this, the negative sign is *stuck to the 2*, no matter what order we add things up in. Go ahead, combine the -2 with the 1. You'll get -1, and now you're left with $7 + (-1) = \mathbf{6}$, the correct answer. Yep, now we don't have to worry about doing the addition and subtraction left to right. Why? Because it's just addition now, and addition has those lovely commutative and associative properties. With addition, we can move the numbers around however we want! Changing subtraction into "adding negatives" is like putting the training wheels on. We're totally safe.

Test Fright

This is such a common issue—it's like stage fright. In acting, many people are fine in rehearsal, but once it's showtime, they freeze up and forget the lines that they've known cold for weeks!

I know this sounds simple, but it's true: Doing well on tests is largely a matter of remaining calm. When you're stressed about something (like an exam), the body produces a physical reaction to the panic you feel: shortness of breath, sweaty palms, sometimes a dizzy or nauseous feeling. Remember that the mind and body work together, so if you can take some slow, deep breaths and mentally imagine your muscles relaxing, you will be able to think more clearly. Relaxing the body really does calm the mind.

Check out the Math Test Survival Guide! on p. 305 so that you can beat Test Fright once and for all!

TESTIMONIAL

Stephanie Perry (New York, NY)

Before: The cool nerd
Today: Finance director for *Essence* magazine!

Growing up, I admit it—I loved school. I loved the sense of accomplishment I felt from finishing my homework and earning good grades. I particularly liked math—I really "got" numbers. I liked that there was always a clear-cut answer to every question.

In high school I kept up my studies, but I also tried to get involved in activities that would get me out of the "nerd" category. I became a track and field manager and a cheerleader . . . so I became known as the "cool nerd." Hey, I was just glad that I wasn't a "pure" nerd.

> **"I liked that there was always a clear-cut answer to every question."**

It's funny—even though I liked math, I still used to think, Why do I need to learn algebra? I'll never use it in the real world!

I was totally wrong.

Today, I'm a finance director for *Essence* magazine. To be a part of this fabulous magazine is a blast! Part of my job involves working with integers, ratios, and various algebraic formulas to stay on top of *Essence*'s financial performance. When it comes down to it, the "numbers" are what keep every business alive. (After all, without sales, no business could stay in business!) When you participate in managing a company's finances, you really feel the importance of what you do. You also get a great perspective on how a business is run, because you learn how much money various departments spend, and how they spend it.

For example, for each issue of the magazine, we have a budget for the advertising pages—meaning, how much companies pay us to run their ads in the magazine (makeup, fashion, you name it). It's important for us to keep track of something called the "variance"—the difference between the real budget and what was *actually* spent. And of course, the variance can either be a positive or a negative integer.

I also love getting to interact with other members of the magazine staff. In addition to having regular consultations with *Essence*'s president and exhilarating conversations about the magazine's financial strategies with the general manager, I've become a trusted confidante for some of the other employees, perhaps because I know the real "secrets" of our business: the numbers. In other words, I know what's *really* spent.

If I didn't have a solid foundation in mathematics, I would never have been able to achieve this level of career success. To all the girls reading this: If you love magazines, but you're not sure you want to be a writer, try what I do—it's great!

What Do You Really Think?

In an anonymous poll, we asked more than 200 girls ages 13 to 18 the following question, and here's what they said.

I'd rather be smart than dumb because . . .

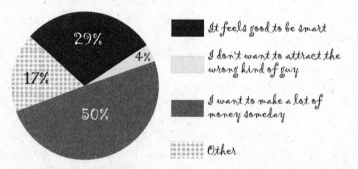

It feels good to be smart

I don't want to attract the wrong kind of guy

I want to make a lot of money someday

Other

Why would *you* rather be smart than dumb? Email me at: notdumb@kissmymath.com, and I'll post some of your answers on the website!

Chapter 3

Mirror, Mirror, on the Wall . . .

Multiplying and Dividing Integers (Including Negative Numbers!)

Do you ever stare at yourself in the mirror? Okay, do you stare at yourself in the mirror *all the time*? Some girls stare in the mirror and like everything they see. But most wish they could change something about how they look, and when they look in the mirror, they obsess about the stuff they *don't* like.

Here's the interesting part: If you can find some things you do like about yourself in the mirror and you focus on those things, you'll start to feel more confident, and you'll actually *become* more attractive to yourself and everyone else, too! Life's too short to focus on what we don't have, anyway. And everyone feels insecure about *something* in that mirror—everyone.

And now here's the strange part about what you see in the mirror: It isn't what other people see, anyway! What you're seeing is a *mirror image* of yourself; everything is flip-flopped. Do this with a friend: Go stand near a mirror, look at your friend in person, then look at her in the mirror, and then look back at her again. You'll be amazed at how subtly, and yet profoundly, *different* she'll look to you—and not just because most mirrors tend to make us all look a little greener. I bet you'll like her more in person. Well, that's how other people see you, too!

Try holding this book up to the mirror. Now hold up a *second* mirror (even a small compact will work), and look at the reflection of the reflection of the book; it is the opposite of the *opposite* of the book. As you probably know, you'll now be able to read the letters again. Believe it or not, all this mirror work makes it easier to understand multiplying and dividing integers . . .

For example: $5 \times -2 = -10$, but $-5 \times -2 = 10$. How can two negatives multiply to give a positive answer? As you might have guessed, it works the same way as holding up two mirrors.

QUICK (REMINDER) NOTE Multiplication

To indicate multiplication between numbers, we can use $\times$, $\cdot$, ( ), or [ ].

Here are some examples:

Some ways to write 5 times 3:

$5 \times 3 = 5 \cdot 3 = (5)(3) = (5)3 = 5(3) = [5][3] = 5[3] = [5]3 = 15$

So the above example could have been written: $(5)(-2) = -10$ and $(-5)(-2) = 10$.

It's good to get used to the parentheses notation. It's used more and more throughout pre-algebra and algebra—maybe because $\times$ looks too much like X!

Multiplying and Dividing Integers . . .
and Other Numbers, Too

Just like images in the mirror are *opposites* of each other, numbers have opposites, too. On the number line, there are numbers on both sides of zero, and for every number like 6 or -2, we can find its *opposite* on the other side.

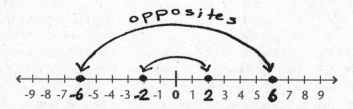

The opposite of 6 is -6, and the opposite of -2 is 2.

What's It Called?

The Opposite of a Number

The *opposite* of a number is its mirror image on the number line; opposites can be positive or negative. For example, the opposite of 2.5 is −2.5. The opposite of −8 is 8. By the way, when you combine (add) opposites together, their sum is always zero. For example, 2.5 + (−2.5) = 0, and −8 + 8 = 0.

Multiplying by −1: Holding Up the Mirror

So, how can you *get* the opposite of any number? Just multiply it by −1. Multiplying by −1 is like saying "the opposite of."

$$-1 \times 5 = -5 \qquad\qquad -1 \times -5 = 5$$

The opposite of 5 is −5. The opposite of −5 is 5.

Or, using parentheses notation: (−1)(5) = −5 and (−1)(−5) = 5.

One negative sign means that there is *one* mirror, and you get the opposite. *Two* negative signs means that there are *two* mirrors, and you'll end up right where you started!

$$(-1)(-1)(5) = 5$$

The opposite of the opposite of 5 is 5

QUICK NOTE When you see something like −7 or −(7), that negative sign is the *same thing* as a −1 multiplying times the 7. So: −7 = −(7) = (−1)(7). They're all different ways of writing the *same thing*: "the *opposite* of 7" which, as you know, is −7.

In fact, the opposite of the *opposite* of every number is the original number! So, for every number n, this is true: $(-1)(-1)n = n$. In other words: $-(-n) = n$.

Division, Too!

Division works the same way: Dividing by -1 has the exact same effect as multiplying by -1. They both mean "the opposite of."* For example, $5 \div -1 = -5$; in other words: $\frac{5}{-1} = -5$. And, dividing two negatives creates a positive (for the same exact reasons), so for example, $\frac{-1}{-2} = \frac{1}{2}$.

Here are the rules for multiplying and dividing positive and negative numbers:

Positive × Positive = Positive	$(+)(+) = +$
Positive ÷ Positive = Positive	$\frac{(+)}{(+)} = +$
Positive × Negative = Negative	$(+)(-) = -$
Positive ÷ Negative = Negative	$\frac{(+)}{(-)} = -$
Negative × Positive = Negative	$(-)(+) = -$
Negative ÷ Positive = Negative	$\frac{(-)}{(+)} = -$
Negative × Negative = Positive	$(-)(-) = +$
Negative ÷ Negative = Positive	$\frac{(-)}{(-)} = +$

* This is because the **reciprocal** of -1 is also -1. (To find the *reciprocal* of a number, write it as a fraction, and then flip it over!) As you may remember from the rules of fraction division, *dividing* by a number (in this case, -1) is the same as multiplying by its reciprocal (in this case, also -1). To review reciprocals, see Chapter 5 in *Math Doesn't Suck*.

Look at the little negative signs in the box above. Every time you see a negative sign in multiplication or division, imagine that a mirror is being held up. If there's no mirror, then you look normal. If there's one mirror, then the image is the opposite. If there are two mirrors, then the image is back to normal again!

And what if you have a bunch of numbers multiplying or dividing together? Here's a shortcut alert to help us: Just count the negative signs!

Shortcut Alert: Counting Negative Signs

In an expression where multiplication and division are the only operations, if you count an *odd* number of negative signs (one, three, five, etc.), the answer will be *negative*. If you count an *even* number of negative signs (none, two, four, etc.), then the answer will be *positive*. Just think of the mirrors!

For example, if we wanted to multiply $(-2)(-4)(3) = ?$, we could count up the negative signs. There are *two* of them total (an *even* number), so the two mirrors cancel each other out, as if there were no mirrors at all, and the answer will be positive: $(-2)(-4)(3) = 24$.

On the other hand, if we wanted to multiply $(-2)(-4)(-3)$, we'd count *three* total negative signs (an *odd* number), which tells us that the answer will be negative: $(-2)(-4)(-3) = -24$.

The same goes for division: Just count the negative signs: $(-9) \div (-3) = ?$ There are two negative signs total, so the answer will be *positive*:

$$(-9) \div (-3) = 3$$

Step By Step

Multiplying and dividing integers:

Step 1. First, make sure you're considering an expression whose *only* operations are multiplication and division.

Step 2. Count the total number of negative signs in the problem. If there's an odd number, the answer will be negative; if there's an even number, it'll be positive.

Step 3. Forget about the negative signs. Multiply (or divide) the numbers as usual, and then, if the answer was supposed to be negative, stick a negative sign on at the end. Done!

*And...
Action!*

Step By Step In Action

Let's do this problem:

$$(-1)(-2)(-3)(-4) = ?$$

Step 1. Yep, this is an expression with only multiplication.

Step 2. We just count up the negatives: There are four of them, an *even* number, so we know our answer will be positive.

Step 3. We multiply the numbers as usual: $(1)(2)(3)(4) = 24$. We know the answer is supposed to be positive, so we're done! Answer: **24**.

Take Two: Another Example

How about a division problem in fraction notation:

$$\frac{(-4)(2)(-5)}{(-2)(5)} = ?$$

Step 1. Yes, this is an expression with only multiplication and division.

Steps 2 and 3. Again, we can just count up the three negatives and then do the problem without the negative signs, *knowing* we'll stick one on later:

$$\frac{(4)(2)(5)}{(2)(5)} = \frac{(4)(2)(\cancel{5})}{(\cancel{2})(\cancel{5})} = 4$$

So the answer is: **−4**.

By the way, if you're afraid you'll forget to stick on the negative sign at the end, you can always just write it like this, putting the negative sign on the outside of parentheses, so that it's "waiting" for the problem to be simplified and then *has* to get stuck on:

$$\frac{(-4)(2)(-5)}{(-2)(5)} \text{ (odd \# of neg's, so)} =$$

$$-\left[\frac{(4)(2)(5)}{(2)(5)}\right] = -\left[\frac{(4)(2)(\cancel{5})}{(\cancel{2})(\cancel{5})}\right] = -[4] = -4$$

Do it whichever way you prefer!

Watch Out!

Don't go canceling negative signs if you have something like this: $-2 - 5$ or $\frac{-2 + y}{-5}$. As soon as addition or subtraction is involved, all bets are off.

QUICK NOTE   When you have a *negative fraction*, you can move the negative sign pretty much anywhere you want. You can even "factor" out the –1 and write it as a positive fraction *times* –1. These all mean the same thing:

$$-\frac{1}{3} = \frac{-1}{3} = \frac{1}{-3} = (-1)\left(\frac{1}{3}\right)$$

And this sort of makes sense—after all, in multiplication and division, it doesn't matter where the negative signs are, just *how many* there are. And in this case, there's always just the one negative sign, right?

Take Three: Yet Another Example

How about something like this: $\frac{(-2)(-3)}{(-4)} + \frac{(-9)}{(-2)} = ?$

Step 1. This expression uses addition, but that doesn't mean we can't apply our steps to each term on its own. Do you see why? We'll just use our counting technique on each fraction, pretending nothing else exists for the moment. Once we've handled each term separately, we can bring together our results and see what happens! So let's do that, starting with the first term, which only uses multiplication and division: $\frac{(-2)(-3)}{(-4)}$.

Step 2. We count a total of three negative signs, so this term will be negative.

Step 3. We can evaluate it by reducing the fraction,* forgetting about the negative signs because we know we'll stick one on at the end:

. . . .

* I also could have used normal "canceling" notation, but I wanted to remind you of something for a moment: Whenever we cancel factors, we're really just dividing the top and bottom by common factors. Do you remember *why* we're allowed to do that without changing the value of the fraction? If not, check out "copycat fractions" in Chapter 6 of *Math Doesn't Suck*.

$\frac{(2)(3) \div \mathbf{2}}{(4) \div \mathbf{2}} = \frac{3}{2}$. So the first term is $-\frac{3}{2}$. At this point, our problem looks like this: $-\frac{3}{2} + \frac{(-9)}{(-2)}$. Better, right?

Now, attacking the second term of the expression, $\frac{(-9)}{(-2)}$, we see that there are two negative signs, so we know they'll cancel each other out to give us $\frac{9}{2}$. This can't be reduced, so there's nothing else for us to do with it. Now, our full problem looks like this: $-\frac{3}{2} + \frac{9}{2} = ?$ It's easier to see what to do if we rewrite $-\frac{3}{2} \rightarrow \frac{-3}{2}$, which we know we can do from the QUICK NOTE on p. 46. Now we've got a common denominator, right? So adding our two terms comes down to regular "combining mints" addition in the numerator: $\frac{-3}{2} + \frac{9}{2} = \frac{-3+9}{2} = \frac{6}{2}$. And now we can reduce: $\frac{6}{2} = 3$.

Answer: $\frac{(-2)(-3)}{(-4)} + \frac{(-9)}{(-2)} = \mathbf{3}$

Doing the Math

Evaluate these expressions. I'll do the first one for you!

1. $\dfrac{-(5)(-2)}{(3)(-5)} - \dfrac{3}{(-9)} = ?$

<u>Working out the solution:</u> Oops, there's subtraction in the expression, so first we'll attack the first term, which only has multiplication and division, and pretend the second fraction doesn't exist! We count three total negative signs, so we can now simplify the fraction, knowing it'll be negative in the end:

$$\frac{-(5)(-2)}{(3)(-5)} = -\left[\frac{(5)(2)}{(3)(5)}\right] = -\left[\frac{(5)(2)}{(3)(5)}\right] = -\frac{2}{3}.$$

Now let's attack the second term: $\frac{3}{(-9)}$. We see only one negative sign, so it will be negative. We can reduce this: $\frac{3}{(-9)} = -\left[\frac{3}{9}\right] = -\left[\frac{1\!\!\!/3}{3\!\!\!/9}\right] = -\frac{1}{3}$. Now our full problem looks like this (watch the negative signs!):

$-\frac{2}{3} - \left(-\frac{1}{3}\right)$, and the two negatives in a row become a plus sign:

$-\frac{2}{3} + \frac{1}{3}$. To combine these, we'll move the negative sign on the first fraction up to its numerator, and then we'll be able to combine them, because they'll have a common denominator:

$\frac{-2}{3} + \frac{1}{3} = \frac{-2 + 1}{3} = \frac{-1}{3} = -\frac{1}{3}$.

<u>Answer:</u> $\frac{-(5)(-2)}{(3)(-5)} - \frac{3}{(-9)} = -\frac{1}{3}$

2. $(-1)(-1)(-1)(-1)(-1)(-1)(7) = ?$

3. $\frac{-(-8)(-4)}{-(2)(6)} = ?$

4. $-\left[\frac{(-1)(-1)(-2)}{(-3)(2)}\right] = ?$

5. $\frac{-(5)(-3)}{(3)(-5)} - \frac{9}{(-9)} = ?$

6. $-\left[\frac{-(5)(-3)}{(3)(-5)} + \frac{\cdot\,9}{(-9)}\right] = ?$ (Hint: Simplify the <u>entire inside</u> before touching that outside negative sign!)

(Answers on p. 319)

What's the Deal?
Canceling Negative Signs

Remember on p. 11 when we learned the "Subtracting Negative Numbers" shortcut, but we didn't know *why* we could change subtracting a negative into adding a positive? I mean, why should it be true that $4 - (-3) = 4 + 3$? Well now, missy, you're ready to find out! For $4 - (-3)$, let's rewrite the subtraction as adding a negative and get this: $4 + [-(-3)]$. With me so far? (First, put your finger over the (-3) and pretend you don't know what's in there, if that makes it easier to follow.) We now know that two negatives multiplied together give you a positive, so we can write $4 + [-(-3)] = 4 + 3$. And voilà!

And now, here's another way to think about why *dividing two negatives results in a positive*. Let's take a look at this fraction: $\frac{(-2)}{(-3)}$. We could rewrite each number by "pulling" out the factor of (-1), like this: $\frac{(-2)}{(-3)} = \frac{(-1)(2)}{(-1)(3)}$, and then we can just "reduce" the fraction by canceling the common factor of (-1): $\frac{(-1)(2)}{(-1)(3)} = \frac{(\cancel{-1})(2)}{(\cancel{-1})(3)} = \frac{2}{3}$. It'll be helpful for you to think of it this way later on in algebra when you start pulling out factors from expressions, so that's why I'm showing you now!

Takeaway Tips

Negative signs can be rewritten as multiplication by (-1). So, $-5 = (-1)(5)$.

Multiplying by (-1) is like saying "the opposite of." So, $(-1)(9) =$ "the opposite of 9."

The opposite of the *opposite* of a number gives you the original number again. So, $-(-9) = 9$. Remember the mirrors!

For any expression whose *only* operations are multiplication and division, just count up the number of negative signs: If it's an odd number, then the answer will be negative. If it's an even number, then the answer will be positive.

Negative fractions can hold their negative sign in their numerator, their denominator, or on the outside. They all mean the same thing!

Danica's Diary

DUMBING OURSELVES DOWN FOR GUYS: WHY IS IT SO TEMPTING?

We've all done some version of it: pretended not to know something, just so a guy could "impress" us, or pretended to need a guy's help when we didn't. All topped off with an innocent smile and a giggle or two. *Ugh.*

Is "dumb" actually sexy? No way! So, what's the deal? Why does this seem to be effective with guys when we know that most guys don't actually want dumb girlfriends?

I think this is how it starts: At some point, we're talking to a guy, and we ask him a question about something he knows more about than we do. And then it happens...He stands a little taller, looks a little prouder, and proceeds to help us—to *show off* what he knows. This makes him feel awesome: full of pride, satisfaction, and power, knowing that he has the ability to be able to help us and impress us.

And if this is a guy we *like*? Well, oh my goodness, we want him to keep feeling like this around us, because then maybe we'll end up being together, right? But you can see how this could lead to us pretending not to know stuff—dumbing ourselves down, and, in extreme cases, pretending to be dependent on the guy for every little thing—just to make him feel good in the moment.

In the meantime, unless he's an idiot himself, he's probably secretly wondering how he got hooked up with such a dim-witted girl. Yikes!

Here's a little equation for you:

Guy showing off his ability + Girl being impressed = Guy is very happy

Sure, we *can* achieve this equation by dumbing ourselves down. The price? Oh, besides the fact that guys don't actually want idiot girlfriends? Well, dumbing ourselves down is a hard habit to break, and it sets us up with a pattern for failure for our whole lives. If you dumb yourself down for a guy (or anyone for that matter), you'll be trapped into keeping up the act, because you won't want to be "found out." Then you'll start feeling bad about yourself, and you won't be sure why.

There's got to be a better way . . . and there *is*.

Unless the guy's a total jerk, then he doesn't want you to be dumb. What he wants is to *feel smart around you*. See the difference?

And come on, ladies, we're not giving the guys very much credit, are we? Are we so egotistical that we think the only way to make a guy feel smart is to dumb *ourselves* down? Guys are smart, talented, and interesting all on their own! Couldn't we be just a little more creative and do *both* things; that is, be smart ourselves and also present guys with genuine opportunities to "show off"?

Try this: Pay attention to the guy you like, and find out a couple of things that he's an expert on— things that you don't know much about. Then, any time you feel tempted to pretend you don't know something, instead find an opportunity to change the topic to something you actually *don't* know much about, and ask him about it! I mean, why not? You might actually learn something, too.

Here's something else: Guys are, generally speaking, physically much stronger than us girls. That's a physiological fact. So, here's another tactic to try: Ask a guy to lift something heavy for you, or open a tight jar, or reach to a high shelf (if he's tall). He might not let on, but this will make him feel great! The more experience you have, the more you'll learn this for yourself, but trust me: Guys always love showing off their physical power. Always. It must be a primal thing.

Listen to me: There is *no need* to dumb yourself down for a guy. The sign of a good relationship is one in which both people learn and grow from each other, so use your conversations with guys as an opportunity to learn new stuff from them—for real!

(And if you know more than he does about *everything*, chances are, he's not a good match for you, right?) The more you do it, the better you'll get, too. The benefits? He'll have more respect for you, and more importantly, you'll have more respect for *yourself* and will be much more able to develop into your full potential.

Write to me, and tell me how this works for *you* at notdumb@kissmymath.com. I love to hear your stories!

Chapter 4

A Relaxing Day at the Spa

Intro to Absolute Value

Who says math needs to be stressful? Why couldn't it be more like going to the spa? Well, now it can be: Welcome to the world of absolute values! This chapter is full of very relaxing, very *positive* things. We'll get to the spa in a moment.

Absolute Value

Remember on p. 5 how we combined the two mint-egers, one negative and one positive, and ended up with 0? That's because 6 and −6 are exact opposites of each other.

$$-6 + 6 = 0$$

So −6 has the same *strength* as 6; that's why they canceled each other out perfectly. **Absolute value** refers to how *strong* the mint-eger is, regardless of whether it's positive or negative; in other words, "absolute value" refers to a number's <u>distance from zero</u>. Even though −6 has a much lesser *value* than 6, they are the same distance to zero (6 steps) on the number line, so 6 and −6 have the same *absolute value*.

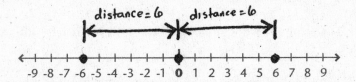

The way we symbolize *absolute value* is with these little bars: | |. So here's how you would write "the absolute value of −6 is 6": |−6| = 6.

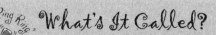

What's It Called?

Absolute Value

The absolute value of a number or expression is its *distance* to zero on the number line. For example, $|6| = 6$, $|-6| = 6$, and $|4 - 7| = |-3| = 3$.

QUICK NOTE   The absolute value of zero is zero!

$$|0| = 0$$

The Absolutely Positively Fabulous Spa

The space inside those little absolute value bars must be a very happy place, like a relaxing spa, because anyone who comes out always ends up being so happy—so *positive*, if you know what I mean. Check it out: $|-163| = 163$. Is she happy? *Absolutely positively,* yes!

In fact, no matter what's inside those bars, <u>absolute values of expressions are *never* negative.</u> Look at it this way: The absolute value of a number is that number's *distance to zero*, and distance can never be negative. After all, you can't run -2 miles, not even if you're running backward.

Step By Step

Evaluating absolute value expressions:

Step 1. First, simplify what's *inside* the bars until you're left with a single number.

Step 2. Remove the negative sign (if the remaining number is negative). Then take off the absolute value bars, and change them into normal parentheses if it helps. Done!

Watch Out!

Don't ever remove the negative signs *before* you've simplified what's inside. For example, for something like $|-5 - 3|$, if you take away the negative sign before simplifying the inside, you might get something like $|-5 - 3| \rightarrow |5 - 3| = 2$. And that would be wrong! If we simplify first, like we should, then we get $|-5 - 3| = |-5 + (-3)|$ (looks like two negative mint-egers getting together to create an even more negative mint-eger!) $= |-8|$. And guess who's getting a massage and maybe a pedicure? Yep: $|-8| = $ **8**.

Those negative signs can be slippery little suckers, so always simplify the insides first, and *then* take off the negative sign (if there's still one left after you've finished simplifying). Above all, when dealing with absolute value problems, don't skip steps like you might in other types of problems. The answers just aren't as straightforward as you think they're going to be . . .

And... Action! Step By Step In Action

Evaluate $|-10 - (-3)|$.

Step 1. First, let's simplify what's inside: $-10 - (-3)$. We know from Chapter 1 that those two negatives in a row can become a plus sign, so we get $-10 + 3$, which is like mints in our mouth: $-10 + 3 = -7$. So our problem now looks like this: $|-7|$.

Step 2. Take off the negative sign and drop the bars. No need for parentheses here: 7.

Answer: $|-10 - (-3)| = |-7| = $ **7**

QUICK NOTE   Just like with parentheses, when a number is right next to the absolute value bars, it means multiplication. So, $3|8| = 3 \times |8|$.

Take Two: Another Example

Evaluate 3|4 − 9|.

Step 1. First, we'll simplify what's inside the bars:
4 − 9 = 4 + (−9) = −5.

Step 2. Now that our problem looks like this, 3|−5|, and there's a single number left inside the bars, we know we can let it come out and be positive! We'll also change the bars into parentheses since we have a 3 hanging around: 3(5) = 15.

QUICK NOTE   For the PEMDAS order of operations (see p. 21), you can think of absolute value bars | | just like you do parentheses ( ) or brackets [ ]. You do whatever's inside them first. After you simplify the inside and make sure that the final number that comes out is positive, the bars can *become* normal parentheses if you need them; for example, if something is multiplying times the outside of the bars like in the example above.

Take Three: Yet Another Example

Evaluate 5 − |−10 − (−3)|.

According to that QUICK NOTE above, the PEMDAS order of operations tells us to first treat the stuff inside the absolute value bars *as if* they were parentheses, so let's first deal with that before we even *think* about the 5. For the moment, let's pretend the whole problem is just |−10 − (−3)|. Hey, we just did that problem back on p. 55, and we found out that |−10 − (−3)| = 7. Yep, positive, because it came out of the spa, didn't it?

Now let's do the *whole* problem:

$$5 - |{-}10 - (-3)|$$
$$= 5 - |{-}7|$$

(Notice that the two negs cannot cancel each other.)

$$= 5 - 7$$
$$= -2$$

And that's the final answer: $5 - |{-}10 - (-3)| = -2$.

I know what you're thinking. A negative answer? In the absolutely positive chapter of spa treatments? What kind of spa is that?

Well, even though what *came out* of the spa was happy, right outside the door, that positive 7 was immediately met with a negative sign. It's okay, she can go back to the spa again tomorrow. (What was 7 thinking, taking a business call on a Sunday?)

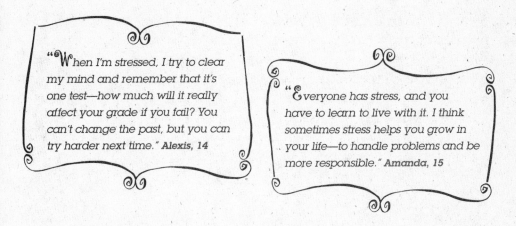

"When I'm stressed, I try to clear my mind and remember that it's one test—how much will it really affect your grade if you fail? You can't change the past, but you can try harder next time." **Alexis, 14**

"Everyone has stress, and you have to learn to live with it. I think sometimes stress helps you grow in your life—to handle problems and be more responsible." **Amanda, 15**

Once more, and I cannot emphasize this enough for all absolute value problems: *Do . . . not . . . skip . . . steps!* If you do, you'll be so, so sorry. *Seriously.* Imagine me staring deeply into your eyes right now, a foreboding look on my face and perhaps an arched eyebrow. Can you imagine it? Okay . . . um, you can stop laughing now.

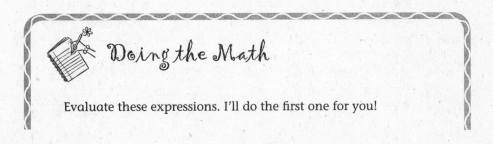

Doing the Math

Evaluate these expressions. I'll do the first one for you!

1. $9 - 2|-4 - (-1)| - 5 = ?$

Working out the solution: First, we'll pretend that nothing else in the world exists except $|-4 - (-1)|$ and figure out its value. Well, *inside* the bars, we can turn the double negative into a positive and get $|-4 + 1| = |-3| = 3$, right? Then, plugging 3 back into our original problem, we get $9 - 2(3) - 5$. Simplifying the multiplication (remember PEMDAS), we get $9 - 6 - 5$. Rewriting the subtraction as "adding negatives" we get $9 + (-6) + (-5)$, and combining left to right, this equals $3 + (-5)$, which equals -2.

Answer: $9 - 2|-4 - (-1)| - 5 = -2$

2. $2|7 - 8| + 5 = ?$

3. $5 - |-3 - (-1)| - 1 = ?$

4. $1 - |-\frac{1}{2}| = ?$

5. $1 - |-\frac{1}{2}| + 2|7 - 8| = ?$

(Answers on p. 319)

Watch Out!

Absolute Values for Variables

I want to point out something: Making numbers become positive when they come out of the absolute value signs only works for *numbers*, not variables! That's because *we don't know* if the value of the variable is positive or negative to begin with. Here's an example. You might be tempted to think that $|-x|$ could be simplified to, say, something like x. But that's wrong! Here's why: Remember, *we don't know* the value of x, which means that for all we know, $x = -5$. And if $x = -5$, then $-x = -(-5) = 5$. And if $-x = 5$, then $|-x| = -x$. Yes, it's possible that $-x$ is a positive number. Pretty crazy, huh? So when you see something like $|-y|$, *never* automatically change it into y, because who knows what $-y$'s distance to zero is? It's *just as likely* that $|-y| = -y$. We just don't know now, do we?

 Takeaway Tips

 Remember, whatever final *number* comes out of the absolute value bars will be positive (or zero). That's because the absolute value refers to the inside quantity's *distance to zero*, and distance is never negative. Or you can think of how *positive* a number might feel after leaving the spa!

Always simplify what's inside the absolute value bars *first*, and then deal with the rest of the problem.

Do not skip steps in evaluating absolute value problems. They are not as straightforward as they sometimes look, and it's easy to make mistakes in them!

What to Do If You Fail a Test

So, you studied, you weren't sure how you did, but then the test gets handed back to you and—yikes! Stomach curls up, blood drains from the face. We've all been there. Yes, me too. Breathe.

It's okay. I'm going to give you a new perspective. First, even though it might feel like the end of the world right now, you've got to ask yourself: How important will this test score seem in 5 years? How about in 10 years? Truthfully, you probably won't even remember it. In fact, the only way that test score could have any sort of lasting negative impact on you is if you *give up on yourself* because of it. And that would be silly!

Your success in life is certainly not determined by any single test score. Believe me, we've all failed a test here and there. (I remember getting a 52% on a math test in college. I'll never forget that orange, handwritten score.) Ups and downs are a *part* of life. What will set you apart and make you special is your ability to bounce back and believe in yourself enough to keep working on improving yourself—and your skills as a student.

In fact, without failure, you'd never have your *belief in yourself* tested, and *that process* can actually make you stronger. Even though it doesn't feel good, it's a natural and important part of life.

It's easy to believe in yourself when things are going great, right? Seems like anyone could do *that*. But are you really going to be a fair-weather friend to yourself? The challenge is to believe in yourself even when you've hit a snag, failed a test, or made some other mistake.

What does it mean to believe in yourself when you've failed? It doesn't mean that you say, "Yes, it's good to fail." Not at all. (See? Now you know I'm not a head case.) What it means is that you actively look for the opportunity to improve. Check out the ideas on p. 305. Maybe you learn that you need more sleep. Maybe you need to finally get help in the tutoring center or online. Remember—it's important to ask for help when you need it!

Also, it's okay that you "had to fail" in order to learn these things. If you read any of those books about incredibly successful businesspeople, they usually failed A LOT before they found their success. The difference is simply how they *handled* the failure. Get stronger, get more determined to succeed, and above all, be *nice* to yourself in the process. What if you had a devastated baby kitten after she tried to walk for the first time and failed? What would you do for her? You'd give her love and then *encourage her to keep trying*.

Treat yourself that nicely too. You need your strength now more than ever! If you beat yourself up, you'll waste the energy that you need to relearn the topics from that test. (Especially in math, make sure you understand topics, even if it's after the test, because you'll need them to understand what's coming next!)

So when you get a bad test score—and it happens to the best of us—look at it as an opportunity to practice the *life skill* of overcoming bad news, learning from it, and looking for ways to move forward, ways to be better *because* of it. And above all— learn to be your own best friend, through thick and thin!

What Do You Really Think?

In an anonymous poll, we asked more than 200 girls ages 13 to 18 the question below, and here's what they said!

Are your math teachers ready to help you?

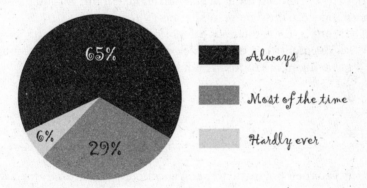

- Always
- Most of the time
- Hardly ever

Surprised?
*Almost all teachers (65% + 29 % = **94%**) are ready to help their students most or all of the time. So even if you're feeling shy, if you've got a question during or after class, go ahead—ask your teacher!*

Chapter 5

Long-Distance Relationships: Are They Worth It?

Mean, Median, Mode

At some point, you may be asked to look at a list of numbers like this:

$$12, 3, -2, 8, 9, 3$$

And you might be asked to find the **mean**, **median**, and **mode** of the list.

Ugh. Just when you thought you had enough math vocabulary words to learn, here are three more, all at once!

The truth is, these methods are not hard to understand. It's keeping the *names* straight that can be tricky. But once you read this chapter, you'll never again be confused about which is which, I promise.

Mean

Let's say your new boyfriend went on vacation with his family for a week, leaving on a Friday. You guys didn't talk on the phone while he was away, but you traded text messages.

You're still not sure how much he likes you. I mean, can you survive the long-distance thing? So maybe you secretly kept track of how many text messages he sent each day in your journal. If you did, then you might have made a list like this:

Friday	Saturday	Sunday	Monday	Tuesday	Wednesday	Thursday
9 texts	7 texts	4 texts	2 texts	0 texts	1 text	5 texts

Here's the list of just the numbers:

$$9, 7, 4, 2, 0, 1, 5$$

Hmm. He must have been really busy on Tuesday and Wednesday, but he's still interested, right? I mean, how many times did he text you per day, on average? To calculate the average, you just add up all the numbers in the list and divide by 7, the total number of days he was gone:

$$\text{Average} = \frac{9 + 7 + 4 + 2 + 0 + 1 + 5}{7} = \frac{28}{7} = \frac{28 \div 7}{7 \div 7} = \frac{4}{1} = 4$$

So he texted you, on average, 4 times a day—not too shabby. Seems like he really likes you, and he's not so bad at this long-distance thing either. He could be a keeper!

What's It Called?

Mean

The **mean** is just the average of all the numbers in a list. It's kind of a strange word for "average," but just think of all that work we have to do to find it, all that adding up and then dividing. That's sort of *mean*, isn't it?

Step By Step

Finding the mean of a list:

Step 1. Combine (add) all the terms in the list.

Step 2. Put that answer over the total number of terms in the list.

Step 3. Simplify the fraction, and you're done!

QUICK NOTE   Be sure to always count any zeros in the list when figuring out the number to divide by!

Find the mean of this list: 4, 1, 0, 6, 4

Step 1. Add up all the terms: $4 + 1 + 0 + 6 + 4 = 15$.

Step 2. Count the total number of terms: 5. Now divide 15 by 5 and get $\frac{15}{5}$.

Step 3. Simplify: $\frac{15}{5} = 3$.

Answer: The mean of this list is **3**. Done!

QUICK NOTE   You might see negative numbers in a list. When this happens, just combine all the terms, positive and negative, and proceed normally. For a review of combining negative numbers, see pp. 5–12.

Take Two: Another Example

Find the mean of these chilly winter temperatures: −4, 11, 0, 15, −9, −1. *These are all in degrees Fahrenheit.*

Step 1. Combine the terms: $-4 + 11 + 0 + 15 + (-9) + (-1)$. First we'll combine $-4 + 11 = 7$, and then we'll combine that with the next term, etc., and get this: $-4 + 11 + 0 + 15 + (-9) + (-1) = \mathbf{12}$.

Step 2. Count the number of terms: 6. Now divide 12 by 6 and get this: $\frac{12}{6}$.

Step 3. Simplify: $\frac{12}{6} = 2$.

Answer: The mean temperature is **2 degrees**. Brrr!

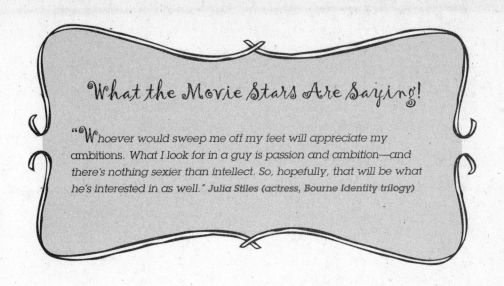

Median

I think the word *median* sounds a lot like *medium*. As in: *Small, medium, large.* So when you get a list and you are asked to give the *median* of the list, first put the list in order so you can see which are the small, medium, and large numbers.

If you were given:

$$-1, 6, -7, 5, 7, 4, 6$$

First, you'd put it in small-*median*-large order. And keep the repeats, if there are any.

$$-7, -1, 4, 5, 6, 6, 7$$

←Small   median   large→

The median will then be the number in the *very center* of the list. Done!

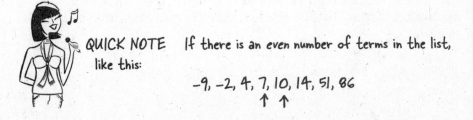

QUICK NOTE   If there is an even number of terms in the list, like this:

$$-9, -2, 4, 7, 10, 14, 51, 86$$

. . . then there is no *single* center term to pick for the median. What do you do?

Just take the two center terms, in this case 7 and 10, and find their average by adding them together and dividing by 2:

Average of 7 and 10 is $\dfrac{7 + 10}{2} = \dfrac{17}{2} = 8.5$

So the median of that list is 8.5.

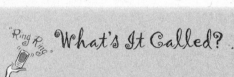

"Ring Ring" **What's It Called?**

Median

The **median** is the number (or the average of two numbers) that sits in the *very center* of the list once you've put the list in small-*median*-large order.

Step By Step

Finding the median of a list:

Step 1. Put the list in order from smallest to largest, keeping all repeats of numbers.

Step 2. If there is an **odd** number of terms in the list, then find the number in the very center of the list. That's the median!

Step 3. Or, if there is an **even** number of terms in the list, take the *two* numbers in the dead center and find their average. Their average is the median!

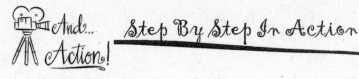

Find the median of this list: −1, 9, 7, −5

Step 1. First, we'll put the list in order: −5, −1, 7, 9.

Step 2. Notice that there is no *single* center term.

Step 3. There are *two* center numbers, −1 and 7, so we'll find their average by adding them and dividing by 2: $\frac{-1+7}{2} = \frac{6}{2} = 3$.

Answer: The median is **3**.

Mode

Let's say you picked up someone's iPod and scrolled through her current playlist. You see a couple of pop songs, a little classic rock, and a ton of samba music. I mean, we're talking, every other song is samba and there's not much else. Since there are more samba songs than any other kind of song, I guess she's in the *mood* for samba, huh? **Mode = mood.** So her "mode" is samba; it happens the most often.

Here's a "playlist" of numbers. What is this list's **mode**?

$$-1, -7, 2, 42, -7, 18, -7$$

Because −7 happens more often than any other number, the **mode** of this list is −7.

QUICK NOTE *More than one mode:* By the way, you can have more than one mode if your list has two (or more) numbers that both appear the same number of times, and more than any other number. For example, the list 1, −6, 4, 5, 1, 1, 4, 6, 6, 4 has two modes, 1 and 4, because they each appear three times and no other number occurs that many times.

No mode: If a list has no repeats (like in the texting boyfriend example), then the list has no mode.

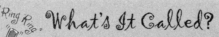

What's It Called?

Mode

The **mode** is the number that appears the most number of times in the list.

Think mode → *mood*. A list can have more than one mode, or no mode at all.

Step By Step

Finding the mode of a list:

Step 1. Count to see which number appears the most and more often than any other number. That's the mode. Sometimes there are two (or more) modes, or there can be no mode. Done!

Step By Step In Action

What's the ***mode*** *of this list:* 3, 1, −4, 6, 3, 8*?*

Step 1. Count and see that 3 appears twice, and all the other numbers appear only once.

Answer: 3 is the mode.

Take Two: Another Example

Find the ***mode*** *of this list*: 6, 4, 0, 6, 7, 4, 6, 4, −8

Step 1. Count and see that 6 and 4 each show up three times, which is more often than any other number.

Answer: The modes are **6** and **4**.

Mode is not so bad, right? Just think "mood," and you'll be fine.

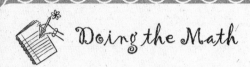 Doing the Math

Find the **mean, median,** and **mode** of each of these lists. I'll do the first one for you.

1. 3, 7, −4, 8, 0, 1, 8, 7

<u>Working out the solution:</u> **Mean:** Okay, this is the one that requires the most work: It's mean, remember? Mean is the same thing as average: In order to find it, we average *all* of the numbers together. So, we add them all up and then divide by the number of terms. There are 8 terms total, so that's what we'll divide by:

$$\frac{0 + 1 + (-4) + 3 + 7 + 7 + 8 + 8}{8} = \frac{30}{8}$$

$$= \frac{30 \div 2}{8 \div 2} = \frac{15}{4} = 4\overline{)15.00}^{3.75} = 3.75$$

So the mean = **3.75**.

Median:
First, let's put the list in small–*median*–large order to find the center term:

$$-4, 0, 1, 3, 7, 7, 8, 8$$

There are an even number of terms, so we need to pick the center two numbers, which are 3 and 7. Now we take their average: $\frac{3 + 7}{2} = \frac{10}{2} = 5$. So the median = **5**.

Mode:
This list is clearly in the "mood" for both 7 and 8. The terms 7 and 8 each appear twice, more than any other number, so there are two modes for this list. Modes = **7** and **8**.

<u>Answer:</u> The mean, median, and modes of this list are **3.75, 5,** and **7 and 8,** respectively.

2. 1, 2, 3, 4

3. 3, −6, 3, 3, 0, 15, 3

4. 15, 2, 7, 2, 7, 3

5. −4, −4, −1, −2, 0, −1

(Answers on p. 319)

 Takeaway Tips

 To find the *mean* of a list of numbers, find the average of *all* the terms by adding them up and then dividing by the number of terms (don't forget to count zeros!). With all of this adding and dividing, finding the mean requires a lot of work, so it's kinda *mean*.

To find the *median* of a list of numbers, think small, med<u>ian</u>, large. Put the numbers in order from least to greatest: The *median* is the one in the very center of the list (or the average of the two in the middle for lists that have an even number of terms).

The *mode* is the number(s) that happen(s) the most times in a list. The list's "mood" = *mode*. A list can have one mode, more than one mode, or no mode at all!

TESTIMONIAL

Jane Davis (New York, NY)

Before: Shy girl with acne and braces
Today: Fashion-savvy financial strategist at Polo Ralph Lauren!

Let's face it: In junior high, everyone has an awkward stage—but I think mine was worse than most. I had a bad haircut, braces, and bad skin. Because of this, I didn't like calling any extra attention to myself. I particularly hated answering questions in class—especially math class. Mostly, I was afraid that if I answered a question wrong, people would make fun of me or think I was stupid.

> "Everyone has an awkward stage— but I think mine was worse than most."

During the summer before my freshman year of high school, a lot of good things happened for me: I got my braces off, my skin improved, and I made the freshman cheerleading squad. I entered high school feeling much more confident. I was still dreading being called on in class—but when I entered my honors freshman math class, almost half of my cheerleading squad was in there, too! These girls were all pretty, popular, AND smart. As the semester began, I watched them raise their hands, eager to participate, and even give *wrong* answers on occasion. And you know what? No one ever made fun of them. It was all in my head. So I began to raise my hand, too—and found that, usually, I knew the right answer after all. I really worked hard in math all through high school, and I'm glad I did! In college, I was admitted to an honors business program, which included several math classes.

After graduating, most of my girlfriends looked for "fun jobs" in marketing, fashion, or advertising. I did the same, but had no idea that

most of those fun jobs require a lot of math!
Eventually, I was hired as an assistant buyer for
Polo Ralph Lauren. That's right—shopping for
clothes was now part of my *job*. Shortly after I
started, the vice president who hired me (who, by
the way, is a fabulous woman and Harvard MBA grad)
told me that it was *because* of my "numbers"
background that I got the job!

As an assistant buyer (in addition to the
shopping!), I used math to analyze sales figures,
determine the appropriate quantities of clothes to
buy, and track orders. I'm now a distribution
analyst/planner, and I use math every day to make
sure stores receive appropriate quantities of each
new style. I'm also responsible for creating sales
reports that the whole company uses!

Part of what I report is something called the
total "average inventory," which is the average
number of units of inventory (clothing, etc.) from a
store over a certain period of time. For example,
the average inventory for one month is just the
average of the store's inventory at the beginning of
the month (BOM) and at the end of the month (EOM).
If someone were to ask me, "On average, how many
units of inventory did we have in our store this
month?" I would find the answer using the formula

$$\frac{BOM + EOM}{2}$$

The average inventory for *six* months, on the other
hand, would require adding up BOMs for each of the
six months, plus adding in the final EOM, like this:

$$\frac{BOM1 + BOM2 + BOM3 + BOM4 + BOM5 + BOM6 + EOM}{7}$$

In the same way that you found the average (or
mean) of a list of numbers in this chapter, I simply
add up all the inventory amounts and then divide by
the sum of these amounts!

Most people don't realize that fashion is a real
business, and behind all the designs are complex
financial plans and strategies. Luckily for me, my
math background prepared me well—I feel so grateful
to have a job that I love!

You Said: Most Embarrassing Moments in School!

Ever been embarrassed at school? Did your face turn beet red? That's okay—it happens to the best of us! Check out these stories, and if you think you have one to put these to shame, email it to: story@kissmymath.com.

"My most embarrassing moment would have to be falling asleep in my geometry class. The night before, I had stayed up really late working on a term paper and had gotten no sleep at all. We had a geometry test that day, and the test was pretty easy. I finished with about a half hour left. I was so tired that I put my head down. Bad choice. That was it for me—I fell into a deep sleep, and even the bell didn't wake me up. I woke up to a dark classroom with everyone gone! I was totally late to my next class—with a drool spot on my shirt! It was pretty embarrassing." **Alex, 15**

"Sophomore year, I was a member of our school's winter guard. We had just learned a new routine called "Mortal Combat," and it was pretty awesome, if I do say so myself. We practiced hundreds of times. The night of the performance I was as excited as ever. I'd performed in front of huge audiences 40 to 50 times, but halfway through the routine, I blanked! The only part I could remember was the end, where I was supposed to drop to the ground and pretend I was dead . . . so I just stood there waiting. As soon as the moment came for me to "die," I fell to the floor with a bang—literally. I don't remember much after that. Apparently, I hit my head pretty hard and had "self-inflicted" whiplash. Eventually, I was able to perform again, but it took a while for the laughter to die down." **Stephanie, 17**

"I didn't used to be the most organized person. You remember the desks where you could put your coloring box, books, and papers inside? I kept so many old assignments wrinkled up and wadded in mine, I had to SIT on my textbooks! One day, my classmates decided I needed to clean out my desk. They flipped my desk upside-down when I wasn't looking! As all those papers came pouring out . . . so did the bugs. Apparently, spiders had decided to make their home in my old homework. All the girls were screaming, and the boys made a game of stomping on them to kill them. Of course, I was crying of embarrassment! Since that day, I have never had a messy desk, and I learned how to throw old papers away!" **Chelsea, 17**

"One night, my algebra teacher assigned an insane amount of homework, but because of my social life, I decided to just not do it and instead go to some school function—a hockey game, I think. I came into class the next day completely unprepared. Of course, this was the day the teacher decided to check my row's homework. She was so upset with me when I claimed that I didn't understand the assignment. I seriously thought she was going to slap me. That hour was the worst hour of my life. Mrs. Hicks put me on the spot and tortured me with questions that I had no ability to answer. I felt so idiotic, sitting there while the other kids tried whispering answers to me. I was so embarrassed." **Nesreen, 15**

"I was so scared to go to high school, because I thought it was just going to be bigger people trying to bully me around. Plus I have the last name "Moran"—just think about that for a second. After I finished registering for my freshman classes and was walking away, one of the teachers called out in a deep, bellowing voice, "Hey Moron, did you steal my pen?" My face turned as red as a cherry, and I wanted to run out of the building, because my best friends, mom, and brother were behind me, laughing away. Yes, he thought he was just being funny, but to hear a man with a deep voice say that and then hear it echoing through the building was horrifying! Every time I saw him that year, he would just look at me and laugh because he knew that I had gotten so embarrassed." **S. Moran, 17**

"At my school, when the bell rings to dismiss us, everyone runs out of the building as if it were on fire, especially on Fridays. One Friday, I was stuck in the crowd on the stairs, and I noticed an extremely cute guy—who smiled at me! Next thing I knew, I did a flip in the air and took everyone on the stairway down with me. About 30 people landed on top of me—and to make matters worse, I landed right on top of the cute guy. I was trying to think of something witty to say, but instead just said, "Well, fancy meeting you here." He just gave me a weird look. Luckily, nobody was hurt, but they were all mad. As I was hurrying away, someone yelled "Hey klutzilla, want your shoe?" When I turned around, the cute guy was holding my right shoe in his hand. I took it but didn't even bother to put it back on—I just got the heck out of there! But guess what? The next day, no one had even heard about my embarrassing moment. I suppose not every stumble on the road of life is as big of a deal as we may think!" **Jessica, 18**

Poll: What Guys Really Think . . . About Smart Girls

We asked more than 200 guys, ages 13 to 18, what they thought about smart girls and girls who dumb themselves down. Here's what they had to say!

What kinds of girls do you like the most?

GUYS SAY:

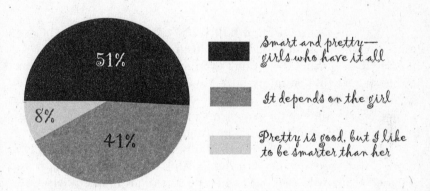

- 51% — Smart and pretty—girls who have it all
- 41% — It depends on the girl
- 8% — Pretty is good, but I like to be smarter than her

As you can see, most guys are either open-minded (and perhaps wary of "show-off" types), or they actively want girls to have it all, which includes being smart.

Notice the 8% of guys who actually prefer to be smarter than the girls they date. You'll want to avoid these guys—and anyone else who doesn't support you being the best you can be.

If you've ever noticed a girl dumbing herself down around you, how did it make you feel?

GUYS SAY:

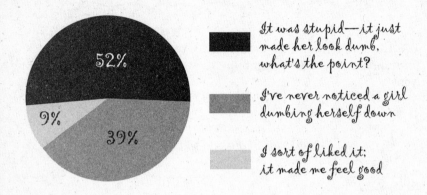

It was stupid—it just made her look dumb, what's the point?

I've never noticed a girl dumbing herself down

I sort of liked it; it made me feel good

As you can see, more than half (52%) of polled guys see "playing dumb" for what it is—a bad idea that just makes girls look stupid!

More than a third, 39%, don't even notice it happening. It's almost hard to believe that so many guys have never noticed a girl dumbing herself down. Assuming you go to a co-ed school, haven't you noticed it at some point? Some of these guys are only 13, but still. I guess we girls are just more observant about some things.

And then there's the 9% who actually feel better about themselves, knowing that girls are playing dumb. This makes me sick! Can you imagine? A guy who *feels good* when he sees a girl making herself less than she is? These are guys to steer WAY clear of. They're likely to be the controlling, domineering type—and someday, possibly even abusive.

For more guys' polls, check out p. 229!

Quiz:
Are You a Stress Case?

*D*oes the tiniest thing make your stomach curl, or are you Ms. Cool?

Take this quiz by expert psychologist Dr. Robyn Landow and contributor Anne Lowney, and see how you fare!

1. It's that time again! Your teacher is handing back last week's math test, and you're pretty nervous. You don't think you did very well. When suspicions are confirmed—you got a D—what happens next?

   a. Your heart sinks, and you start to worry that your low score will pull down your overall grade in the class. Maybe later you'll think of ways to get your grade up, but for now, you just wish you could forget about it.

   b. You're disappointed that you didn't get a good grade and realize you'll have to spend more time studying for the next test. You're determined; maybe you'll sign up for tutoring hours after all.

   c. Panic. You feel like you might start crying at any moment. No matter what you try to tell yourself, it feels like the end of the world.

2. Your worst nightmare is pretty much coming true. Your history teacher has randomly assigned partners for your latest project, and, of course, you end up with the guy who just broke up with you last week! I mean, you only shared one kiss, but your pride is still hurt, and now your face is feeling hot and flushed. How do you deal?

   a. When you hear your two names called together, you immediately look down, trying to think. You don't even look up from your desk to acknowledge him. You're busy plotting all the possible ways to complete the joint assignment without having to actually interact.

   b. You can't do this! It would be way too embarrassing, and it must be avoided at all costs. You wait for class to be over and then go talk to your teacher, making up a story about you and your family going out of town, so you have to do the assignment alone and, um, she better reassign your partner.

   c. Resolve yourself to the fact that this is definitely going to be uncomfortable, but you just have to face it. Maybe you'll make a joke to break the tension or let him know from the beginning that it doesn't have to be weird—you'll just do the assignment and get it over with.

3. We all get a bad teacher now and then, and your science teacher is making your life a nightmare. It seems like she calls you out on your mistakes more than anyone else in the class. She even called you "stupid" yesterday! What do you do?

 a. It's become too stressful to even go to class. For now, you'll use any excuse to get out of it—a trip to the nurse's office for a fake stomachache, an appointment with the guidance counselor—anything to avoid "the monster." If you get called into the principal's office, well, maybe you'll finally break down in tears about the whole thing.

 b. You're too afraid to talk to her; you'd rather pretend the problem doesn't exist. You decide to just keep your eyes down in class and hope she doesn't call on you. Your doodles of her wearing a mustache make you feel better.

 c. You finally decide to tackle the issue—calmly. You let your parents know about the problem and then bravely approach your teacher after class and ask what her expectations are and how, in her opinion, you can improve. As hard as it may be, you make sure to be polite and respectful the whole time. You want to work this out.

4. You've been assigned a chapter of your French book to read for an important quiz tomorrow. Unfortunately, the more you read, the less you understand, and you're feeling stressed! You're sure you'll be *totally* lost when it's time to take the quiz. It's not fair—here you are, really making an effort, right? You feel like crying. What do you do?

 a. You slow down and start over. You can handle this! You break the chapter into parts from the beginning and only move on when you've looked up every word in your French/English dictionary.

 b. You read the same passages over and over. You should know this stuff! You get so frustrated with how much time you're spending on a single chapter that you completely lose your focus.

 c. You can't concentrate at all. Your anxiety flares way up, and eventually you decide it's not worth all the stress. You close the book for good.

5. Your parents keep telling you they expect a better report card than last term. You hate hearing the same thing over and over; it's only making things harder for you. You:

 a. Crack under the pressure and blow up at them when they mention school. You're ready to forget about your grades—just to get back at them.

 b. Try to put their pestering out of your mind and concentrate on the goals you've set for *yourself*.

c. Worry constantly about getting bad grades. Every assignment is like a test, and you find yourself getting tense over small things.

6. Two of your really good friends are in a huge fight. They're both convinced the other one is wrong, and frankly, you can see both sides of the issue. They're both right *and* wrong, but no matter what you say, each of them is only happy if you'll dis the other one, and now they're both turning on you, accusing you of not being a loyal friend! How do you respond?

a. The anxiety of having two friends turning against you because of an argument you had *nothing* to do with is overwhelming! You feel torn, and it seems easier just to side with one friend rather than have everyone mad at you.

b. Naturally you're upset, because you feel like you can't support one friend without betraying the other. So, you decide to stay true to yourself, *not* take sides, and encourage them both to resolve their differences. By keeping your calm, you can stay out of it, and hopefully they'll come back around once they've made up.

c. The situation is impossible! You're getting major pressure from both girls. You don't know what to do except pretend to take both of their sides. You know you're being totally two-faced, but anything is better than getting outcast.

7. It's five minutes before test time. Half of you is saying "relax" while the other half is reviewing your math notes like crazy. You:

a. Take a deep breath and, as you calmly review your notes, remind yourself that you're well prepared and you know this stuff. Anyway, all you can do is your best work!

b. Begin to full-on panic! Your mind races: "What if I didn't study enough? What if I don't get a single problem?" You're half convinced you already failed, and you prepare for the worst.

c. Feel your heart start to speed up and hope you studied the right material. But a quick last-minute review couldn't hurt. Of course, with your heart beating so fast, you're not exactly absorbing the stuff on the review page in front of you.

8. All the math tests are passed out and the teacher gives the go-ahead to start. Your first reaction is:

a. You skim the test to see what you're up against, taking note of the problems you know will be easy. Then you go back and read the instructions carefully. Now you're ready to begin.

b. You race to start the first question. It's a tough one, and you automatically think you're not prepared. This determines your mood through the rest of the test.

c. Here comes that familiar queasy feeling. You studied, but now your mind is totally blank! You don't seem to recognize any of the problems, and the only thing that stands out is the clock above the chalkboard. Tick. Tick. Tick.

9. This time, you've *really* overdone it in terms of extracurricular activities. What were you thinking when you decided you could juggle field hockey, violin, student council, *and* drama club? Who knows? And, especially now that your mom has asked you to help out around the house more, it's pretty clear that you've bitten off way more than you can chew. You:

a. Totally panic! You feel intense pressure to take on many activities, and you can't think of sacrificing any of them. You get so frustrated that sometimes you start crying out of sheer exhaustion or anxiety. Everything is a mess, and you can't seem to fix it. You're paralyzed!

b. You decide something needs to change to free up your schedule, but it seems like you're running from one activity to the next and you never get around to actually making that change. It's easier not to face it.

c. Try to stay calm and not let yourself be overwhelmed. First of all, you create an escape plan from drama club; it's the least important activity for you at the moment. Second, you take a good look at your schedule and think carefully about how you will budget your time. You could spend a little less time on Facebook, after all!

10. You have an English paper due at the end of the week, but you hardly understand the topic and definitely need help getting started. You:

a. Feel uneasy at the thought of talking to your teacher, so instead of asking for help, you just wing it and hope for the best. You work on it for a few days, so you figure it's gotta be at least a B.

b. Avoid the thought of the upcoming due date altogether. The night before, you throw something together . . . that you know won't get a good grade. At least that dark cloud over your head has finally passed!

c. Schedule time with your teacher right away. Maybe she'll help you with your outline. You'll be sure to prepare for the meeting so she won't think you're trying to get her to do your work for you.

11. Somehow you got the dates confused, and you've just realized that you have two tests on the same day—tomorrow! How do you spend the evening?

a. It's stressful enough preparing for one test, much less two. You decide to blow off math and spend all your time getting ready for your history test. Maybe there'll be time to cram in some math later.

b. You start by making a list of the things you need to get done and how much time each will take. You know from experience that if you make a good plan from the start, you won't get overwhelmed.

c. You don't know where to start! Your thoughts keep wandering from task to task, and with all that worrying going on, you get almost nothing done until 10 P.M.! And then you're stressed about having to choose between sleep and studying. Yikes!

Scoring

1. a = 2; b = 1; c = 3   **5.** a = 3; b = 1; c = 2   **9.** a = 3; b = 2; c = 1

2. a = 2; b = 3; c = 1   **6.** a = 2; b = 1; c = 3   **10.** a = 2; b = 3; c = 1

3. a = 3; b = 2; c = 1   **7.** a = 1; b = 3; c = 2   **11.** a = 2; b = 1; c = 3

4. a = 1; b = 2; c = 3   **8.** a = 1; b = 2; c = 3

If you scored between 11–17

Wow, you're as calm as they come! You don't sweat the small stuff, or the big stuff for that matter. Your friends can count on you to stay unruffled even when everyone around you may be losing it. When it comes to challenging schoolwork, you know how to stay relaxed, and you use constructive problem solving to resolve disappointments. Bravo!

That said, if you find that your feathers are *never* ruffled, even when something really disappointing or unexpected happens, you might want to stop and think about *why* you don't seem to care. Be careful not to mistake being carefree for being too afraid to care about things that ultimately matter to you. What are you good at? What are your goals? Often, people hesitate to set goals for themselves because they're afraid to fail. But then they sell themselves short! You might benefit from striving a little more in your life, and you might just be surprised at how good it feels and at what you might accomplish along the way. Remember: Occasional failure is not the end of the world; it happens to all of us! See p. 59 for more on this.

If you scored between 18–27

Congratulations! You've got things pretty much under control. You take your responsibilities seriously and have a lot on your plate, and that's admirable. Even though your worrying can interfere, you do have some basic tools to handle stress. Here are some more ideas.

Try to relax when that stressful feeling creeps up on you. Stretching or exercise helps, or even deep breathing. A good laugh can also decompress most high-stress situations. And if you're having trouble finding something to laugh about, seek out a friend who is always good for a laugh, or watch some cute kitten videos online! And don't forget to enjoy some healthy escapes from the daily grind of school. Take your dog out for a long walk or take an afternoon hike with some girlfriends to get back to nature.

Remember, it's okay if there are little "bumps" with your schoolwork or other things. Everyone has good days and bad days. What's important is that you're doing your best, so feel good about it!

If you scored between 28–33

Everybody gets stressed out sometimes, but it sounds like you stress and worry more than most. You might have too much going on and need to come up with a way of freeing up your schedule. Let's face it, it's difficult to do *anything* well when you are feeling overwhelmed. There's nothing wrong with admitting that you are overcommitted and need more downtime. It might just be the best thing you've ever done for your grades—and your well-being. Perspective is everything, so make a concerted effort to feel good about what you do accomplish. How would others compliment you? What nice things would they say? Well, say those things to yourself—as often as you can! Someone once said, "Worry is a misuse of imagination." Worrying *never* helps. Doing your best is all you can ever ask of yourself. *Says Danica, "When I used to get really stressed out in school, in addition to talking to my parents and teachers, I used to sing a song inside my head called, 'Don't Worry, Be Happy,' by Bobby McFerrin. For some reason that always helped me feel better . . . sometimes it still does!"* Also, if you find that you are losing sleep, getting frequent headaches, or experiencing shortness of breath, it's very important to seek immediate help in learning how to cope with your stress. Teachers, advisers, and even your parents can relate to the stress you are going through and help you work through these nerve-racking situations. It's amazing how much calmer you'll feel when you know you have someone in your corner.

See pp. 159–60 for other great stress-relieving tips, pp. 59–61 for how to deal with failing a test, and p. 305 for the Math Test Survival Guide!—a great way to get rid of pretest jitters.

Chapter 6

The Blind Date
Getting Cozy with Variables

By the end of the next few chapters, you'll be solid in your understanding of how variables work and what they're good for; they won't be such a mystery anymore. Won't that be nice? To accomplish this, we're not going to *solve* for *x* in this chapter. Nope! We're just going to get *cozy* with *x*.

But first, a bigger challenge: You're about to go on your first blind date. You know nothing about him, except that he's your best friend's cousin and he just moved to town. Sure, your best friend will be coming too, but still. You have no idea what to expect. She said you'd totally love him. But will he be charming? Smart? Cute? What will his *value* be?

In math, when we don't know the *value* of something, we can use a **variable** to stand in for that value.* Variables act like placeholders. For some reason, "*x*" is a very popular placeholder. I'm not sure why, but I'm pretty sure it has nothing to do with the fact that *x* is the international symbol for "kiss."

Anyway, back to the guy. So if you really like him, there's a good chance the night will end with you saying to your friend: "He's awesome! Omigod, he's so incredibly *x*! Do you think he'll call me? He's your cousin, so help me out, will you?" Hey, there's gotta be *some* advantage to dating your BFF's cousin.

Since we don't know this guy's value yet, that's why we used *x* to stand in for what you'd say. Later on, you'll be able to fill that part in. His value might be that he's really smart or charming or just simply irresistible.

We'll come back to the guy in a minute. For now, it's time to get something straight: What *is* a variable? And why can they be so scary?

• • • • • • • • • •

* For a full definition of **variable**, see p. 88.

$$X \leftarrow \text{variable}$$

Hmm. I don't know; it doesn't look so scary all by itself like that, does it?

A variable is just a letter that's standing in for a number. The only reason it's a letter is because we don't yet know *which* number it's standing in for; we don't know its *value* yet.

You know, the only reason everyone uses letters like *x* for variables is because of a lack of imagination. You could use whatever symbol you want! Instead of *x*, you could draw a box, or a flower, or even a smiley face.

That's looking much better.

Substituting for *x*

Okay, back to the guy. Remember how we said that after your blind date, the sentence "Omigod, he's so incredibly *x*!" could go a few different ways? Because *x* can stand for different *values*, the expression can have different meanings:

*Omigod, he's so incredibly **x**!*

Let's say **x** = cute. Then "*Omigod, he's so incredibly **x**!*" would become:

→ *Omigod, he's so incredibly **cute**!*

If **x** = funny, then "*Omigod, he's so incredibly **x**!*" would become:

→ *Omigod, he's so incredibly **funny**!*

Let **x** = smart, and then "*Omigod, he's so incredibly **x**!*" would become:

→ *Omigod, he's so incredibly **smart**!*

In the same way, if someone gives you the expression $3x + 5$, then you can substitute numbers for *x*, and figure out the expression's new value.

Let's say $x = \mathbf{2}$. Then $3x + 5$ would become:

$$\rightarrow 3(\mathbf{2}) + 5 = 6 + 5 = \underline{11}$$

What if $x = \mathbf{0}$, then $3x + 5$ would become:

$$\rightarrow 3(\mathbf{0}) + 5 = 0 + 5 = \underline{5}$$

If $x = \mathbf{10}$, then $3x + 5$ would become:

$$\rightarrow 3(\mathbf{10}) + 5 = 30 + 5 = \underline{35}$$

If $x = \mathbf{-1}$, then $3x + 5$ would become:

$$\rightarrow 3(\mathbf{-1}) + 5 = -3 + 5 = \underline{2}$$

> ♪ **QUICK NOTE** It doesn't matter if the problem uses x, y, z, or any other letter—like a, b, or c—or even something wacky like j. They're all just placeholders.
>
> When you are given a number to substitute for a variable, I highly recommend using *parentheses* to surround the new number (like I did above). It'll help you keep things straight later on, and it's a great habit to get into now!

Consider this expression: $7 + 2a$

What would this expression equal if $a = 3$? What if $a = -3$? What if $a = \text{❀}$? What if $a = 3b$? Let's start plugging stuff in and find out!

If $a = 3$, then the expression becomes $7 + 2(3) = 7 + 6 = \mathbf{13}$.

If $a = -3$, then the expression becomes $7 + 2(-3) = 7 + (-6) = \mathbf{1}$.

If $a = \text{❀}$, then the expression becomes $7 + 2(\text{❀}) = \mathbf{7} + \mathbf{2}\text{❀}$. There's nothing else we can do!

If $a = 3b$, then the expression becomes $7 + 2(3b) = \mathbf{7} + \mathbf{6b}$. Again, there is nothing else we can do.

Remember: *Whatever* value you're given, just stick it in wherever the original variable was, and you'll be fine!

> ♪ **QUICK NOTE** When you are asked to substitute values for x (or any other variable), you might see that part of the question written a few different ways:
>
> If $x = 5$ Let $x = 5$ When $x = 5$ Where $x = 5$
> For $x = 5$ Let's say $x = 5$

These all say the *same thing*. They all mean, "Let's say that *x* has the value of 5. Now go ahead and use 5 wherever you see *x*." Then simplify to get your final answer. That's it!

Sometimes, instead of using the word *substitute*, textbooks might ask you to *evaluate* expressions, like this: Evaluate $3x + 5$, when $x = 4$.

This is the same thing we've been doing: Just stick **4** in for *x*, and then simplify:

$$3x + 5 \rightarrow 3(4) + 5 = 17$$

We've found the value of $3x + 5$ when $x = 4$. In other words, for $x = 4$, we **evaluated** the expression $3x + 5$: It equals 17.

Ring Ring. What's It Called?

Evaluate

To *evaluate* means to "find the value of," which is easy to remember because it has the word *value* in it. Well I guess it only has *valu* in it, but let's not go splitting hairs here, okay? We usually *evaluate* expressions by substituting (plugging in) a number where there used to be a variable.

Example: *Evaluate* the expression $2w - 5$ for the value $w = 1$.

Then we'd do $2(1) - 5 \rightarrow 2 - 5 \rightarrow 2 + (-5) = $ **−3**.

Doing the Math

Evaluate the following expressions by substituting the values I give you. I'll do the first one for you.

1. Evaluate $2y + 1$ for each of these y values: $y = 0$, $y = 2$, $y = -5$, and $y = \frac{1}{4}$.

Working out the solution: To find the expression's value when $y = 0$, we substitute 0 wherever we see y, being sure to use parentheses. So the expression would be:

$$2(0) + 1 = 0 + 1 = 1$$

If $y = 2$, we stick in 2 wherever we see y. So:

$$2(2) + 1 = 4 + 1 = 5$$

If $y = -5$, we substitute -5 wherever we see y. So:

$$2(-5) + 1 = -10 + 1 = -9$$

If $y = \frac{1}{4}$, we stick in $\frac{1}{4}$ wherever we see y. So:

$$2\left(\frac{1}{4}\right) + 1 = \frac{2}{1}\left(\frac{1}{4}\right) + 1 = \frac{2 \times 1}{1 \times 4} + 1 = \frac{2}{4} + 1 =$$

(reducing the fraction) $\frac{2 \div 2}{4 \div 2} + 1 = \frac{1}{2} + 1 = 1\frac{1}{2}$

Answers: $1, 5, -9, 1\frac{1}{2}$

2. Evaluate $4 + 3g$ for each of these g values: $g = \begin{smallmatrix}\text{❀}\end{smallmatrix}$, $g = 1$, $g = -1$, $g = \frac{1}{3}$, $g = 0.2$.

3. Evaluate $2h + \frac{6}{h}$ for each of these h values: $h = 1$, $h = 2$, $h = -3$, $h = \odot$.

(Answers on p. 319)

Now that we've brought more art into the classroom, I thought we'd clear up some math vocabulary once and for all. This isn't going to be as boring as you'd think, I promise. (Remember, if this were boring, I would have stopped writing, and you wouldn't be holding a book right now, would you?)

Parts of a Math Expression

What Are They Called?

Let's take a look at a little math expression and name its parts:

$$3x + 5$$

Coefficient → · ↑variable · ↖ constant

Variable: This is a letter like *x* or *y* that stands for a value you don't know. It could have a *variety* of different values; hence the name "variable"—for, um, variety. Personally, I like to think of *x* as a bag of pearls; we just don't happen to know *how many* pearls are in the bag. (In equations, when you solve for *x*, you *find out* how many pearls are in the bag.*)

$$x = \text{🝾}$$

Also, while you're still getting comfortable with this stuff, feel free to write in a little ☐ or ⚘ wherever you see *x*; it tends to make things easier to understand at first. (If you're really artistic, you could even draw a 🝾!)

$$3x + 5$$

Coefficient → · ↑variable · ↖ constant

Constant: This is something that will make you feel all warm and mushy inside. You know what's constant? The sun. Yes, the sun rises every morning. Even if it's cloudy, you know the sun is out there somewhere. It's a *constant* in life, unchanging. Makes me feel all warm and fuzzy just thinking about it.[†]

In the expression $3x + 5$, the number 5 is just the number 5. Constant. How nice. On the other hand, who can really trust *x*? We don't know much about *x*; its value could be anything. We

* * * * * * * * *

* We'll talk more about solving for *x*, and bags of pearls in Chapters 7 and 12!
† Technically, the sun doesn't really "rise" so much as the Earth rotates at a rate of more than 1,000 miles per hour, and in a few billion years, the sun will likely expand into a red giant and probably engulf the Earth, disintegrating all remaining life. But I'm going for a warm and fuzzy vibe here, okay?

might get to find out sometime, through doing an algebra problem like $3x + 5 = 11$ and solving for x; then again, we may *never* know. When you see something like 5, however, you think of the good old days when we didn't deal with things like letters in our math homework—just nice, predictable, "constant" numbers.

(By the way, the 3 in $3x$ is *not* considered a constant because it's a secret ally to that mysterious x. It will make the mysterious value 3 times what it used to be. But we still don't know how big $3x$ is, because we don't yet know how big x is! And that's not particularly warm and fuzzy, now, is it?)

So, when you see a number all by itself with *no* variables stuck to it, and you're feeling all nice and comforted, you know you've got a *constant*!

$$\underset{\text{Coefficient}}{3}\,\underset{\text{variable}}{x} + \underset{\text{constant}}{5}$$

Coefficient:* Yes, this is the secret ally to the variable that we were just talking about. For the expression $3x + 5$, the coefficient is the 3. It's the number that is "stuck" to the variable with multiplication. (Remember: $3x$ means $3 \times x$.) But how can we remember the name *coefficient*?

Well, $3x$ is a more *efficient* way of writing $3 \times x$, right? (And it's a much more efficient way of writing $x + x + x$.) So you might be tempted to call the 3 an "Efficient."

But how can we remember that it's called a <u>co</u>efficient? Check this out: 3 is a number, which at first might look like a constant, right? So, feeling relieved, you might think, "Oh, good, it's a <u>co</u>-," but then you stop dead in your tracks, noticing that it's stuck to a secretive variable, and you correct yourself—"-Efficient." And maybe that's just how it got its name: coefficient.

. . . .

* FYI, some textbooks call this a numerical coefficient.

Variables, coefficients, and constants all get together to make up the *terms* in math expressions. As we learn tricks that deal with variables in the following chapters, we'll keep talking about terms and what it means to deal with just one term of an expression. For example, $3x + 5$ has two *terms*. One term is $3x$ and the other is 5.

Let's nail down this definition, too!

"Ring Ring" What's It Called?

Terms

Terms are numbers and/or variables that are stuck together with multiplication and division. For example, 7 is a single term. So is $\frac{8xy^2}{3}$, because it's all stuck together like that. Single terms are separated from each other by addition or subtraction. For example:

$$xyz + 7a - 4a + \frac{x}{2}$$

. . . has four terms. See how each *term* is made up of things that are schmooshed together? Here's how I like to think of it: If you're close enough to spread a *germ*, you're part of the same *term*!

Next we'll go over different kinds of coefficients and constants: fractions, decimals, and yes, even negative numbers. But first, one more word about terms for the interested few.

What's the Deal?

(This is an optional section for very inquisitive minds only!)
Consider an expression like this: $3(x + y)$. How many *terms* does it have?

That's a very good question. Technically, it has only one term, because you've basically just got "two things" being *multiplied* together: 3 and $(x + y)$. Sure, there's addition in there, but only <u>inside</u> the parentheses. In Chapter 10, you'll learn that you can distribute the 3 to get: $3(x + y) = 3x + 3y$. At that point, you'd have two terms, because the addition is separating the $3x$ from the $3y$. But while it still looks like this, $3(x + y)$, there's only one term. Because it's a single term, it's considered a product of *factors*. The two factors* are 3 and $(x + y)$.

Why do we care so much about what a "term" is? Because in later chapters (and throughout algebra), there are times when rules apply <u>only</u> to single terms. I mean, it's not totally pointless or I wouldn't tell you about it, I *promise*.

QUICK NOTE   We always write $3y$, not $y3$. Technically, they mean the same thing, but always write the coefficient first. Seriously, everybody does it, and this is one popular trend worth following!

Fractions and Decimals

Constants and coefficients can be fractions and decimals, too. For example:

$$\frac{2}{3}x + 0.09$$

In this expression, the constant is 0.09 and x's coefficient is $\frac{2}{3}$.

. . . .

* In Chapter 1 of *Math Doesn't Suck*, I teach factors that are *numbers*. (The factors of 6 are 2 and 3, for example, because they multiply to give you 6.) For *variables*, factors work the same way. In the expression $10y$, two factors are 10 and y, because they multiply together to give you $10y$. We'll see variable factors again on p. 114.

Here's a cool factoid for you: When the coefficient is a fraction, it can be written two different ways. The variable can either be on the outside of the fraction, or it can be in the numerator. They are two ways of writing the exact same thing! For example:

$$\frac{3}{4}x = \frac{3x}{4} \text{ and } \frac{1}{2}z = \frac{z}{2}$$

(Remember: $1z = z$.)

I'll talk more about *why* this is true on p. 118. The reason I'm telling you this now is because you might see an expression like this:

$$\frac{5y}{2} + 8$$

What is y's coefficient, after all? Hmm. Is it 5, or is it $\frac{5}{2}$? Well, if the coefficient is 5, then what is 2? We know that 2 can't be a constant, because it's stuck to the 5 and the y; it's not separated from the y with addition or subtraction like the 8 is.

The whole thing becomes *much* clearer when you remember that $\frac{5y}{2}$ can be <u>rewritten</u> as $\frac{5}{2}y$. What a relief! And now you can see that y's coefficient is actually $\frac{5}{2}$. Pretty nifty, eh?

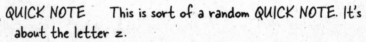

QUICK NOTE   This is sort of a random QUICK NOTE. It's about the letter z.

Can I even just tell you how many times I've ended up with strange (and totally wrong) answers, because I couldn't tell the difference between 2 and z in my handwriting? Eventually I started writing my z's like this: Ƶ, and I highly recommend it! You'll thank me later.

(It's also not a bad idea to draw a line through your 7's like this 7̶, so you don't confuse them with your 1's. And I'm sure I don't have to tell you that the letter o makes a terrible variable, but if you're ever forced to use it, then just be careful. You can always draw a line through your zeros like this ⌀.)

Watch Out!

We know that $\frac{3}{4}x$ and $\frac{3x}{4}$ mean the same thing. But if we see the x on the *bottom* of the fraction, it's an entirely different expression altogether. Just try plugging the value $x = 4$ into these three expressions: $\frac{3}{4}x$, $\frac{3x}{4}$, and $\frac{3}{4x}$. You'll get an answer of 3 for the first two, but an answer of $\frac{3}{16}$ for the last one.

Here's a really helpful trick for counting money that uses *decimal numbers* as coefficients. Check it out!

Reality Math
Coins in Your Change Purse!

Have you ever had a bunch of coins in your change purse and wanted to know how much money is there? Sure, you could just add it all up, coin by coin. *Let's see: A quarter is $0.25, plus a nickel—that's $0.05, plus—oh, there's another quarter, etc.* But with lots of coins, this could take forever! I want to show you a trick, and once you get the concept, it's so much faster. Plus, some word problems require this same method, so it'll help with homework, too.

First, let's imagine that you had only quarters. If you had 3 quarters, then the total money from quarters would be ($0.25) $\times$ 3 = $0.75, right? If you had 6 quarters, you'd have a total of ($0.25) $\times$ 6 = $1.50. If you had 10 quarters, you'd have a total of ($0.25) $\times$ 10 = $2.50. Notice a pattern? Looks like we could say that the total amount of <u>money</u> from quarters can be found by multiplying ($0.25) $\times$ q, where q stands for the *number of quarters* you have. We can also write the multiplication like this: ($0.25)q. This means you can just stick the *number of quarters* into this formula, ($0.25)q, and you'll know the total amount of money you have from those quarters. Notice that there are two separate values here: the *number* of quarters (the variable), and the total amount of <u>money</u> from the quarters (what you get when you multiply the variable times 0.25). We can get from one to the other via this simple formula: ($0.25)q. Are you with me so far? (If not, just read this paragraph again. You'll get it!)

As you can imagine, dimes, nickels, and pennies all work the same way. If the *number* of dimes is d, then the total amount of money from dimes would be ($0.10)d, because each dime is worth $0.10. If the *number* of nickels is n, then the total amount of money from the nickels would be ($0.05)n, because each nickel is worth $0.05. The same goes for pennies, where p is the number of pennies: ($0.01)p.

Now we can put it all together. Just count the *number* of quarters, dimes, nickels, and pennies you have and substitute them into this formula:

Total money from your change purse =

$$(\$0.25)\mathbf{q} \;+\; (\$0.10)\mathbf{d} \;+\; (\$0.05)\mathbf{n} \;+\; (\$0.01)\mathbf{p}$$

where $\mathbf{q}$ = the *number* of quarters, $\mathbf{d}$ = the *number* of dimes, $\mathbf{n}$ = the *number* of nickels, and $\mathbf{p}$ = the *number* of pennies.

So empty your change purse and get out your calculator. You can figure out how much money you have quickly and easily!

"When you don't understand something, don't be afraid to ask a question. You're probably not the only one who doesn't understand, and remember this: You look more stupid with an F on your paper than you do asking for some help during a lesson." **Cody, 14**

"Ask for help even if you think you look dumb. Ask, because you will be the smart one in the long run. I promise!" **Lana, 15**

Negative Constants and Coefficients

Constants and coefficients can be *negative numbers*, too. But these are a little trickier.

See, if you have an expression like $4y - 5x - 2$, then technically, the constant term is -2, not 2, and x's coefficient is actually -5. Just like with negative numbers, when you see subtraction in an expression that has variables in it, the safest bet is to turn ALL the subtraction into "adding negatives." Check it out:

$$4y - 5x - 2 \rightarrow 4y + (-5x) + (-2)$$

And now it's easier to see why x's coefficient is -5 and the constant is -2.

Changing subtraction into "adding negatives" is so incredibly helpful, because, let's face it, negative numbers can be confusing *wherever* they pop up!

Watch Out!

We know that when we see a variable with no coefficient, its coefficient is actually 1, right? Well, sometimes we'll *subtract* a variable that seems to have no coefficient, like this: $2 - x$. According to the text above, we should rewrite this as $2 + (-x)$, and now we can see that x's coefficient is actually $-\textbf{1}$. Pretty sneaky, huh? This is just another reason to always rewrite subtraction as "adding negatives!"

The sneaky coefficients 1 and -1 are often lurking around unseen, but you can write them in if it helps. So, for $x - y + 4z$, you would first rewrite subtraction as "adding a negative": $\rightarrow x + (-y) + 4z$.

And then you could write in the sneaky coefficients: $1x + (-1y) + 4z$.

Constants are much nicer. If you don't see a constant, that means it's zero, which is kind of like not having a constant, right? It's much more straightforward. But then again, that's why constants make us feel so nice and warm inside.

Doing the Math

For each of the following expressions, say how many terms there are, and then name the variables, constants, and coefficients. Be sure to rewrite subtraction as "adding negatives." I'll do the first one for you.

1. $(0.6)g - 8 - h + \dfrac{11}{12}$

<u>Working out the solution:</u> First, we'll rewrite the subtraction as "adding negatives": $(0.6)g + (-8) + (-h) + \dfrac{11}{12}$. Now let's count terms; it looks like there are four (all the things now separated by + signs). The variables are easy—just look for the letters. That's g and h. The constants are also easy; they're the numbers all by themselves, so that's -8 and $\dfrac{11}{12}$. What are the coefficients? Just look to see what's stuck to the variables. So, g's coefficient is 0.6, and h's coefficient is -1.

<u>Answer:</u> There are **four terms total**; the variables are g and h; the constants are -8 and $\dfrac{11}{12}$; the coefficients are 0.6 and -1.

2. $7 - 4z$

3. $n - m$

4. $0.2 + a - 5b + \dfrac{2}{3}c$

5. $\dfrac{3x}{5} - 9 - y$

(Answers on p. 319)

Takeaway Tips

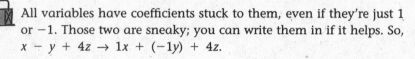 Variables are placeholders for numbers *whose value we don't know yet*, but sometimes we plug in (or substitute) actual numbers for them. This is called substitution.

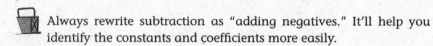 Always rewrite subtraction as "adding negatives." It'll help you identify the constants and coefficients more easily.

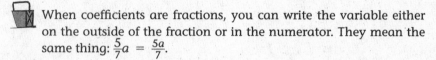 All variables have coefficients stuck to them, even if they're just 1 or −1. Those two are sneaky; you can write them in if it helps. So, $x - y + 4z \rightarrow 1x + (-1y) + 4z$.

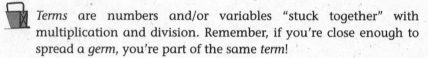 When coefficients are fractions, you can write the variable either on the outside of the fraction or in the numerator. They mean the same thing: $\frac{5}{7}a = \frac{5a}{7}$.

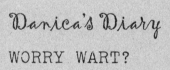 *Terms* are numbers and/or variables "stuck together" with multiplication and division. Remember, if you're close enough to spread a *germ*, you're part of the same *term*!

Danica's Diary

WORRY WART?

Generally speaking, I was a good student in junior high. But I used to spend so much time *worrying* about getting bad grades. I'd be afraid to start studying for a math test because I was afraid I wouldn't do well. (See p. 305 for test-taking tips!) And of course, the more I worried, the more time I wasted not studying. Then I would worry about *that*, which is funny if you think about it: worrying about worrying!

When I let all of this worrying get to me, the whole thing would end with me getting teary-eyed and my mom trying to assure me it wasn't really that important. Sure, doing well on a math test wasn't

going to feed a starving child or bring peace to the
Middle East...but it sure seemed important to me.
I felt that it determined whether I was a success or
a failure. How could that not be important? I didn't
realize that my mom was right: Worrying itself was
my worst enemy. If I let myself relax, things would
get better. Eventually that did happen, but boy did
I waste a lot of time stressing for no reason.

Listen to me now: Worrying only wastes energy,
and it never helps. When you feel stressed or
worried, take slow, deep breaths and think of
something that makes you smile: a pet, a song, a
joke, anything to break the tension and lighten your
mood. See pp. 159-60 for great ideas for stress
relief!

Backpack Too Heavy?
Adding and Subtracting with Variables

*I*f you've ever had trouble figuring out why $3h - h$ doesn't equal 3, why $3h - 3$ doesn't equal h, or what the heck $h - 3h$ equals, then it's time to get a few things straight, once and for all!

By the way, we're going to be using the word *coefficient* all over the place in this chapter. So if you haven't already, check out the box on pp. 88–9.

Backpack . . . Too . . . Heavy

I remember having the heaviest backpack when I was in junior high. And at the time, it wasn't cool to have a traditional backpack with two straps, so I carried all my books in a book bag that slung over just one shoulder—my right shoulder. By the time I got into high school, I'd figured out how to use my locker more effectively, so I carried a lighter load, but I kid you not: Today, if I'm standing up perfectly straight, my right shoulder is *still* lower than my left shoulder!

The thing is, I'm sure I didn't need to be carrying around everything in that bag. Heck, I bet there were half-full bottles of water, granola bars, hair clips I thought I'd lost, and other items I didn't even realize were in there. I didn't know the full contents of my bag, and yet I carried the stuff around and did things with that bag all day long.

Similarly, when we "do things" with variables, we're handling "bags" of unknown quantities. We don't know their value, but we can still *do things* with them!

Bags of Pearls (Variables) and
Simplifying Expressions

As I mentioned on p. 88, I look at x as a bag that holds some unknown number of pearls. Remember, we may not know *how many* pearls are in each bag, but that doesn't have to stop us from carrying them around and doing things with them. (It doesn't stop us with our book bag, now, does it?)

Here's an expression involving a variable and addition:

$$3 \; + \; 2a$$

3 pearls   2 bags of pearls

We can't combine or simplify this any more than it already is. It's two bags of pearls and 3 loose pearls: What more can I say? However, if instead we had:

$$3a \; + \; 2a$$

3 bags of pearls   2 bags of pearls

. . . then we could add those 3 bags to the 2 bags and get a total of 5 bags of pearls:

$$3a + 2a = 5a$$

Makes sense, right? In the same way, if we have $8x - 2$, then we're saying that we have 8 bags of pearls minus 2 loose pearls. There's nothing else we can do to simplify that expression. But, if we instead had:

$$8x - 2x$$

. . . then we could subtract 2 bags from the 8 bags and end up with 6 bags of pearls. And $6x$ is a much simpler way of saying $8x - 2x$, don't you agree? That's why we can say that we've *simplified* it.

$$8x - 2x = 6x$$

You might have noticed that in the previous problems, we basically just added or subtracted the <u>coefficients</u> to get the answer:

$$3a + 2a = (\mathbf{3 + 2})a = \mathbf{5}a$$

and

$$8x - 2x = (\mathbf{8 - 2})x = \mathbf{6}x$$

We didn't even need to think about the bags. Sounds like it's time for a SHORTCUT ALERT!

Shortcut Alert

If you want to add or subtract two terms that have the exact same variable part, then you can simply add and subtract their coefficients. Just remember to keep the variable stuck to the answer!

Watch Out!

If you see two terms like $2y$ and $3xy$ in your textbook, notice that they do NOT have the exact same variable part because of the x in $3xy$. So don't try to combine them by adding coefficients! (We'll deal with this more in Chapter 9, when we "combine like terms.")

QUICK (REMINDER) NOTE   Remember that $x = 1x$
This comes in handy when you're combining coefficients, and you can always write the 1 to make it easier.

Step By Step

Adding and subtracting terms with variables in them, using our coefficients shortcut:

Step 1. Make sure the terms that you want to add or subtract have the same *variable* part.

Step 2. Add or subtract (combine) the coefficients.

Step 3. Make sure the variable is still part of your final answer. Done!

Watch Out!

The coefficients shortcut works only for addition and subtraction, NOT multiplication or division. Remember: $6x + 4x = 10x$ can be done by thinking $6 + 4 = 10$ *only* because we have 6 bags plus 4 bags, which is equal to 10 bags.
It's harder to use the bags analogy when we multiply or divide two variables: $(6x)(4x) = 24x^2$. (See the WATCH OUT on p. 109 for more on this.) But if you keep the bags analogy in the back of your mind during addition and subtraction, you won't misuse the shortcut.

And... Action! Step By Step In Action

Let's simplify $7n - 5n$, using the coefficient shortcut.

Step 1. Both terms have the same variable part, n, so we can subtract them.

Steps 2 and 3. Subtract the coefficients: $7 - 5 = 2$. Then stick the n on it: $2n$. And that makes sense because 7 bags minus 5 bags should be 2 bags, right?

Answer: $7n - 5n = $ **2n**

Take Two: Another Example

Let's simplify $5n - 7n$.

Step 1. Yep, both terms have the same variable part: n.

Steps 2 and 3. Using our shortcut, we know that we can just subtract (combine) the coefficients: $5n - 7n = (5 - 7)n$. And we know from combining integers (see p. 5) that $5 - 7 \rightarrow 5 + (-7) = -2$. In life, you can't really have -2 bags; that's why knowing how to use the shortcut is so important.

Answer: $5n - 7n = \mathbf{-2n}$

"Some girls are smart, but they think they need to play it down to be accepted. They will also take rude comments from boys because they think that it is okay to be talked to in that way. So maybe they're really not that smart after all. Truly smart girls—who don't hide it—are important, because there are not enough girls in high-up positions in the world. It's also great to see women in other countries such as Iraq stepping up and trying to improve the world."

Mariel, 16

Take Three: Yet Another Example

Let's simplify this: $9 - (-5x) - 2x$.

Before we start the steps, let's make this look, um, better. I don't like all those negatives, do you? First, let's rewrite the double negative into a $+$ sign, and then let's change the subtraction into "adding a negative": $9 + 5x + (-2x)$. So far, so good? Make sure you remember why this is the same exact expression as above, just written differently. (If you don't remember, try rereading Chapter 1.)

Step 1. The second two terms have the same variable part, x, so yep, we can combine those two: $5x + (-2x)$. Let's do it!

Steps 2 and 3. Using our coefficient shortcut, we can simplify this by combining coefficients; $5 + (-2) = 3$, so we know that $5x + (-2x) = 3x$. Now our full expression is $9 + 3x$. Done!

Answer: $9 - (-5x) - 2x = \mathbf{9 + 3x}$

Watch Out!

At first glance, some people might think that $3h - h$ equals 3. But don't be fooled; even though it looks oddly right, it's totally wrong! Remember that with variables like x or h (or any other letter), we're dealing with *bags of pearls*, and $3h - h$ means "3 bags minus 1 bag," leaving us with 2 bags: $2h$. Using our shortcut, we could remember that h is the same as $1h$. We have $3h - 1h$, and subtracting the coefficients $(3 - 1 = 2)$, we find out that $3h - h = 2h$.

Keep reminding yourself that variables stand for *bags of unknown quantities* (of pearls or whatever), and you'll use the steps correctly!

By the way, on the first page of this chapter, we saw $3h - 3$. As you now know, there's no way to simplify this further: It's just 3 bags minus 3 pearls. And now you can also do $h - 3h = -2h$, because you can just combine the coefficients: $1 - 3 = -2$.

QUICK NOTE   Just like we combine *integers*, we can also combine three or more *variable* terms; we do them two at a time. For example, to simplify the expression $5z - 9z + 6z$, I would rewrite this as $5z + (-9z) + 6z$ and go from there, combining the first two and then adding the third term. Try this one on your own! (BTW, the answer will be $2z$.)

 *Doing the Math*

Simplify these expressions. I'll do the first one for you.

1. $\frac{1}{2}x - \frac{5}{2}x - (-x) = ?$

<u>Working out the solution</u>: First things first: Let's rewrite the double negative as a $+$ sign and change the subtraction to "adding negatives." While we're at it, let's stick in the 1 as the last coefficient: $\frac{1}{2}x + \left(-\frac{5}{2}x\right) + 1x$. So far, so good? Now, combining coefficients, we should do $\frac{1}{2} + \left(-\frac{5}{2}\right) + 1$. Once we move the negative sign to the numerator (see p. 46 to review this), this is just good 'ol fraction addition, but with negative numbers. To combine fractions with the same denominator, we just combine across the numerator, right?
So: $\frac{1}{2} + \frac{-5}{2} = \frac{1 + (-5)}{2} = \frac{-4}{2} = -2$. Now we can combine the -2 with 1, and we get $-2 + 1 = -1$. What did this all mean again? Oh yeah, we're combining x's coefficients so we can combine: $\frac{1}{2}x + \left(-\frac{5}{2}x\right) + 1x$. The sum of the coefficients is -1. This means that our answer is $-1x$, which is the same as $-x$.

<u>Answer</u>: $\frac{1}{2}x - \frac{5}{2}x - (-x) = -x$

2. $9j + 3j - 5j = ?$

3. $11c - 4c - (-7c) = ?$

4. $0.8y - (-0.3y) - 0.9y = ?$

5. $\frac{1}{2}z - \frac{1}{4}z = ?$

6. $7t - 2t - (-t) + 10 = ?$ *(Hint: Keep that constant separate from the variable terms.)*

(Answers on p. 319)

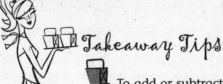

 To add or subtract terms with the same *variable* terms, simply add or subtract their coefficients. Just be sure to keep the variable stuck to it.

When simplifying an expression, it's almost always a huge help to *deal with negative signs first*. If you are adding or subtracting terms, change all the double negatives to positives, and then change subtraction to "adding negatives."

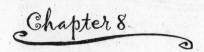

Chapter 8

Something Just Went "Squish"

Multiplying and Dividing with Variables

$\mathcal{N}$ow that we've mastered addition and subtraction, we can move on to multiplication and division. You'd almost think you were back in elementary school, except, oh yeah—the variables! But multiplying and dividing with variables isn't so bad, really, especially if you like squishing things.

Multiplication with Variables—SQUISH!

The good news about multiplying with variables is that you can basically just squish everything together, coefficients *and* variables, and get your answer. You'll see what I mean in a minute.

First, let's review some multiplication symbol stuff, like we did on p. 40, because for whatever reason, there are a billion* ways to write multiplication. It can get especially confusing once variables get involved.

· · · · · · · · · ·

* Of course this is an exaggeration; it's more like eight or nine. Don't start with me, okay?

Multiplication Box with Variables!

For multiplication, we can use ×, •, ( ), [ ], or sometimes *no symbol at all!* Here are some examples.

Ways to express 5 times 3:

$$5 \times 3 = 5 \cdot 3 = (5)(3) = 5(3)$$
$$= (5)3 = [5][3] = 5[3] = [5]3$$

Ways to express 5 times x:

$$5 \times x = 5 \cdot x = (5)(x) = 5(x)$$
$$= (5)x = [5][x] = 5[x] = [5]x = 5x$$

Notice that you can only use "no symbol at all" for multiplication *when a variable is involved.* For instance, $3y = 3 \times y$, and $ab = a \times b$. When two variables are right next to each other, it means they're being multiplied together.

Just as a reminder, this obviously doesn't work for numbers: $34 \neq 3 \times 4$. Similarly, you already know that $2\frac{5}{8} \neq 2 \times \frac{5}{8}$. Remember that $2\frac{5}{8} = 2 + \frac{5}{8}$. For mixed numbers, putting a number right next to a fraction means *addition.**

QUICK NOTE As I mentioned on p. 91, the coefficient always appears *before* the variable, so y • 4 = **4y**, not y4. Also, you'll tend to see letters listed alphabetically. So, you'd see **3ab**, not 3ba. They mean the same thing†, but this is just the way it's done. (And it'll be easier to check your answers in the back of your textbooks if your answers *look* the same!)

. . . .

* This is part of why it's so helpful to convert mixed numbers into improper fractions before "doing stuff" with them. For more on mixed numbers, see Chapter 4 of *Math Doesn't Suck.*
† Remember the commutative property of multiplication from p. 31? Well, this explains why $3xy$ is the same as $3yx$ or $y(3)x$. It just means that the *order* of multiplication doesn't affect its value.

So, taking all of this into consideration, let's say we want to multiply $5x \cdot 4y$. First, we move the 4 up to the front, where it will join the 5 to become the coefficient: $5 \cdot 4 \cdot x \cdot y$. Then, we just multiply and squish everything together: $5 \cdot 4 \cdot x \cdot y = 20xy$.

Watch Out!

If you see something like $7a \cdot 3a$, when you squish the two a's together, you get a^2, not aa. (So $7a \cdot 3a = \mathbf{21a^2}$.) We'll go over exponents in Chapters 15 and 16. I just wanted to show you this one example now.

What's the Deal?

If it's obvious to you that $5 \cdot 3x = 15x$, then you don't really need this box, but you might find it interesting! So, why *does* $5 \cdot 3x$ equal $15x$? It's a fair question. Think of it this way: If you have 3 bags of pearls and then you have 5 *times* that amount, then you have 15 total bags of pearls, right? Or you can think of it as 5 *rows* of 3 bags each, which again, equals 15 total bags.

Also, you could first rewrite the multiplication with parentheses and then use the associative property of multiplication from p. 24, to change the grouping:

$$5 \cdot 3x = (5)(3x) = (5 \times 3)(x) = (15)(x) = 15x$$

So, instead of grouping the 3 and the x, we grouped the 5 with the 3, and then we could multiply them to get 15. It's just another way to look at it!

What the Movie Stars Are Saying!

"I loved school so much that most of my classmates considered me a dork." **Natalie Portman (actress, Star Wars: Prequel Trilogy)**

QUICK NOTE Negative Signs

When multiplying (or dividing) with variables, negative signs work the same way they did with integers in Chapter 3. When you're multiplying, just count the negative signs. If an even number of negative signs are being multiplied together, they will cancel each other away. However, if the number of negative signs is *odd*, you'll end up with a negative sign in the product. That's all there is to it!

Step By Step

Multiplying with variables:

Step 1. Notice whether the problem has any negative signs. If so, count 'em up. Just like with integers, an *even* number of negative signs means they will cancel away, and an *odd* number of negative signs means the product will have a negative sign left over.

Step 2. Move the numbers to the front of the variable(s), and multiply the numbers together to get the new coefficient.

Step 3. Squish everything else together. If the same variable appears more than once, be sure to give it the correct exponent (more on this in Chapter 16). Also, it's best to write the variables in alphabetical order. Done!

Let's multiply $-9x(8y)\left(-\frac{1}{4}\right)$. Don't worry. It's not as bad as it looks!

Step 1. There are two negative signs being multiplied together, so we know they will cancel away. Let's get rid of them now. Hey, why not?

Steps 2 and 3. We move all numbers to the front and squish the rest together: $9(8)\left(\frac{1}{4}\right)xy$. Multiplying the numbers, we can see that the 8 and the $\frac{1}{4}$ will cancel to result in a 2, so the new coefficient will be $9(2) = 18$. That's it!

Answer: $-9x(8y)\left(-\frac{1}{4}\right) = \textbf{18}\textbf{\textit{xy}}$

Watch Out!

If you have something like $5 + x \cdot 4$, make sure that you don't multiply the 4 times the 5. They are separated by addition! $5 + x \cdot 4 = \textbf{5 + 4}\textbf{\textit{x}}$, and that's all you can do with it.

Take Two: Another Example

Let's simplify $(-3)(-x)(-2) - 4x$.

First, notice that *not all of the terms* are being multiplied together. Only the first three terms are being multiplied; the fourth term is being subtracted. Sometimes it can be tricky when a lot of negative signs are hanging around. (The way we know that the first three terms are being multiplied is because their parentheses are touching each other.)

As dictated by the Panda PEMDAS rule (see p. 21), first we'll focus only on the terms being *multiplied*: $(-3)(-x)(-2)$.

Step 1. We count three negative signs being multiplied, so we know we'll end up with a negative sign in the product. Let's rewrite it as $-(3)(x)(2)$.

Steps 2 and 3. We bring the numbers to the front and multiply them, and because there's only one variable, there's nothing else to squish. We're done with the multiplication part of this problem: $-6x$.

Now, let's look at what our full expression has become: $-6x - 4x$. Changing the subtraction to "adding a negative," we get $-6x + (-4x) = -10x$. And we're done!

Answer: $(-3)(-x)(-2) - 4x = \mathbf{-10x}$

 Doing the Math

Simplify these expressions. I'll do the first one for you!

1. $(-7y)(-2)(-x)(-y) = ?$

<u>Working out the solution:</u> Everything is being multiplied together, so everything gets included in the squishing! We count four negative signs, so they'll cancel away completely: From this point forward, we can pretend they're not even there. How nice. Next, we move the numbers to the front to create our new coefficient and squish the rest together in alphabetical order: $(7)(2)yxy = 14xy^2$.

<u>Answer:</u> $(-7y)(-2)(-x)(-y) = 14xy^2$

2. $(8g)(-2gh) = ?$

3. $(-9a)(-5b)\left(\frac{1}{9}a\right) = ?$

4. $(10w)(0.1)(2w) = ?*$

5. $(163v)(0)v(6x) = ?$

(Answers on pp. 319–20)

What's the Deal?
Negative Signs on Variables

You may or may not have noticed something peculiar: Since we've been working with variables, when we count and cancel negative signs, I haven't been referring to their products as being positive or negative. I've only said whether or not the product would have a negative *sign*. The reason is because, frankly, we don't know if the answer is positive or negative; it all depends on the value of x, which we don't know! See the WATCH OUT on p. 58 for more on this. While it's true that the opposite of x is $-x$, and the opposite of $-x$ is x, we still have no idea which one is positive and which one is negative. (Or x could be zero.)

With numbers, if we simplify $-(-5)$, the negative signs cancel, and we know our answer will be positive: 5. On the other hand, with variables, if you see two *negative signs*, the two *signs* still cancel each other out: $-(-y) = y$; we just don't happen to know if y itself is positive or negative, and that's okay. Our job is to get the *signs* right. That's all we need to do.

. . . .

* To review decimal multiplication, see p. 123 in *Math Doesn't Suck!*

Dividing with Variables—In Other Words, *Fractions* with Variables

From now on, you're pretty much only going to see division happening in the form of fractions. Remember, <u>fractions *are* division</u>. The expression $3 \div 4$ means the *same thing* as the expression $\frac{3}{4}$. After all, they will both deliver the same answer, 0.75, right? In the same way, it's true that $y \div 4$ means the same thing as $\frac{y}{4}$.

The main thing to realize is that you can treat variables in fractions pretty much the same way you would treat any other number, because variables basically *are* numbers! We just don't know *which* numbers they are yet.

If you were given these fractions to multiply, $\frac{3}{2} \times \frac{5}{3}$, you'd know what to do, right?* You'd multiply across the top and bottom, and then reduce it by canceling common factors. (Remember, *common factors* refers to the factors that the top and bottom have in *common*.)

$$\frac{3}{2} \times \frac{5}{3} = \frac{3 \times 5}{2 \times 3} = \frac{\cancel{3} \times 5}{2 \times \cancel{3}} = \frac{5}{2}$$

In the same exact way, if you are given these fractions to multiply, $\frac{a}{2} \times \frac{5}{a}$, you'd multiply across the top and bottom:

$$\frac{a}{2} \times \frac{5}{a} = \frac{a \times 5}{2 \times a} = \frac{\cancel{a} \times 5}{2 \times \cancel{a}} = \frac{5}{2}$$

. . . and then you'd cancel factors, just like we would with numbers. If you see an *a* on the top and bottom, it's *some* number, and even though we don't know *which* number it is[†], it's certainly a common factor on the top and bottom, so it can be canceled!

. . . .

* To review reducing fractions by canceling (or dividing by) common factors, see Chapter 6 in *Math Doesn't Suck*.

† Though it better not be zero. More on that in a moment.

Let's review some properties of division; in other words, some properties of *fractions*:

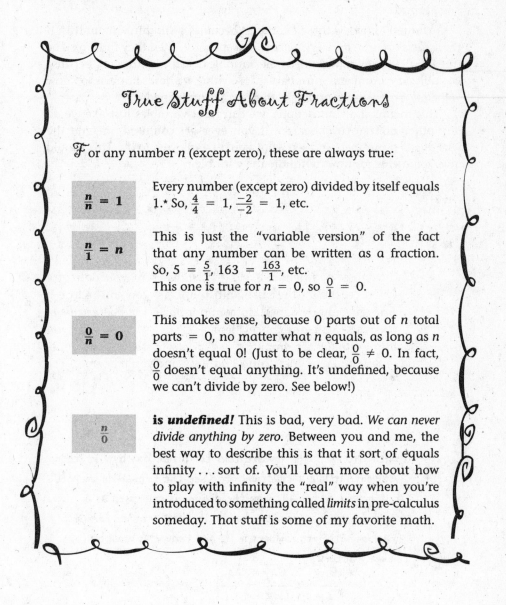

True Stuff About Fractions

*F*or any number *n* (except zero), these are always true:

$$\frac{n}{n} = 1$$

Every number (except zero) divided by itself equals 1.* So, $\frac{4}{4} = 1$, $\frac{-2}{-2} = 1$, etc.

$$\frac{n}{1} = n$$

This is just the "variable version" of the fact that any number can be written as a fraction. So, $5 = \frac{5}{1}$, $163 = \frac{163}{1}$, etc.
This one is true for $n = 0$, so $\frac{0}{1} = 0$.

$$\frac{0}{n} = 0$$

This makes sense, because 0 parts out of *n* total parts = 0, no matter what *n* equals, as long as *n* doesn't equal 0! (Just to be clear, $\frac{0}{0} \neq 0$. In fact, $\frac{0}{0}$ doesn't equal anything. It's undefined, because we can't divide by zero. See below!)

$$\frac{n}{0}$$

is *undefined!* This is bad, very bad. *We can never divide anything by zero.* Between you and me, the best way to describe this is that it sort of equals infinity . . . sort of. You'll learn more about how to play with infinity the "real" way when you're introduced to something called *limits* in pre-calculus someday. That stuff is some of my favorite math.

. . . .

* If you read *Math Doesn't Suck*, you probably remember copycat fractions from p. 64. The top and bottom are copycats of each other, and copycat fractions always equal 1.

In the previous box, when I say n is any nonzero "number," I mean *any* nonzero value, so n could equal 5.6, $\frac{1}{2}$, or -0.4. Check it out:* $\frac{5.6}{5.6} = 1$, $\frac{\frac{1}{2}}{1} = \frac{1}{2}$, and $\frac{0}{-0.4} = 0$.

I wanted to review these examples because, well, think about this: If these facts are true for *every number* (except zero), then they must also be true for all variables (as long as the variable doesn't equal zero), because variables *are* numbers—numbers whose value we don't happen to know yet. So, $\frac{x}{x} = 1$, $\frac{x}{1} = x$, and $\frac{0}{x} = 0$. Also, $\frac{x}{0}$ is undefined.

This means that in fractions, we can treat variables just like we do any other nonzero number. So, if you ever get confused, pretend the variable is a number, and see what you'd be allowed to do. This should give you a clue for how to handle the variable.

Watch Out!

If you see a fraction like $\frac{3 + d}{9 + d}$ make sure you don't try to cancel the d's! Just like with numbers, you can reduce a fraction only by canceling common *factors of multiplication*. So, while you *can* cancel here $\frac{2 \times c}{9 \times c} = \frac{2c}{9c} = \frac{2\cancel{c}}{9\cancel{c}} = \frac{2}{9}$, you cannot do *anything* to further simplify $\frac{3 + d}{9 + d}$. It has to stay the way it is. Just pay attention, don't get "cancel happy," and you'll be fine!

QUICK NOTE   If you are told to "divide $8xy$ by $2y$," then just write it like a fraction, $\frac{8xy}{2y}$ (See the $\frac{3}{4}$ example on p. 114 if it helps.) After all, once the division is expressed as a fraction, you know you can handle it *like any other normal fraction* with just numbers in it, and reduce as usual by canceling common factors:

$$\frac{8xy}{2y} = \frac{\overset{4}{\cancel{8}}xy}{\underset{1}{\cancel{2}}y} = \frac{4x}{1} = 4x$$

Pretty crazy stuff, but I know you can handle it!

• • • • • • • • •

* For a review of complex fractions, see Chapter 9 in *Math Doesn't Suck*.

Step By Step

Multiplying and dividing with variables—squish!

Step 1. Make sure any division is written as fractions. Also, make sure you're only considering one term at a time, whose *only* operations are multiplication and division.

Step 2. Count the total number of negative signs in the product. If there are an even number of negative signs, then they will cancel away. If there are an odd number of negative signs, the product *will* have a negative sign left over.

Step 3. Bring the *numbers* to the front and multiply/divide them.

Step 4. Similarly, multiply (squish!) the variables together, and cancel any common variable factors on the top and bottom of the fraction, just like you would with numbers. As always, it's nice to leave variables in alphabetical order. Done!

 And... *Action!* ## Step By Step In Action

Simplify this expression: $\dfrac{(9n)(-m)}{2m} \div \dfrac{(-5)}{m}$.

This might look scary, but it's not hard when you take it one step at a time!

Step 1. Let's write the division as fractions. Okay, I know what you're thinking: "But this expression has division AND fractions!" Think back to when you learned to divide fractions.* All we did was flip the second fraction and then multiply, and we got an equivalent expression, right? We can do the same thing here:

$$\frac{(9n)(-m)}{2m} \div \frac{(-5)}{m} = \frac{(9n)(-m)}{2m} \times \frac{m}{(-5)}$$

Now, just like normal fraction multiplication, we can multiply across the top and bottom:

$$\frac{(9n)(-m)}{2m} \times \frac{m}{(-5)} = \frac{(9n)(-m)(m)}{2m(-5)}$$

· · · · · · · · · · ·

* For a review of fraction division, see p. 56 in *Math Doesn't Suck!*

Okay, now that all the division looks like fractions, and we have just one term stuck together with multiplication and division, we're ready to continue to the next steps!

Steps 2 and 3. We count two negatives, so the negative signs will cancel away, and we can drop them! Also, let's move the numbers to the front, on the top and bottom of the fraction:

$$\frac{(9n)(-m)(m)}{2m(-5)} = \frac{(9n)(m)(m)}{2(5)m} = \frac{(9n)(m)(m)}{10m}$$

There are no numerical factors to cancel, but we can cancel an m from the top and bottom.

Step 4. Cancel and squish: $\frac{(9n)(m)(\cancel{m})}{10\cancel{m}} = \frac{9nm}{10}$. Finally, we'll put our answer in alphabetical order.

Answer: $\frac{(9n)(-m)}{2m} \div \frac{(-5)}{m} = \mathbf{\frac{9mn}{10}}$

QUICK NOTE   Back on p. 92, I told you that if a variable's coefficient is a fraction, there are two different ways to write it. For example:

$$\frac{3}{4}x = \frac{3x}{4} \text{ and } \frac{1}{2}z = \frac{z}{2}$$

And here's why: If you see $\frac{3}{4}x$, you know it means $\frac{3}{4} \times x$, right? Well, you also know from fraction multiplication (and from the box on p. 115), that you can always write numbers as fractions by giving them a denominator of 1. So:

$$\frac{3}{4}x = \frac{3}{4} \times x = \frac{3}{4} \times \frac{x}{1} = \frac{3 \times x}{4 \times 1} = \frac{3x}{4}$$

And voilà! We've just shown that $\frac{3}{4}x = \frac{3x}{4}$. Try it with $\frac{1}{2}z$ on your own.

Doing the Math

Simplify these expressions, just like we did with numbers on pp. 47–8. (Assume that all variables in the denominators of fractions do not equal zero.) I'll do the first one for you!

1. $(-1)(-a)(-b) + \dfrac{-a}{(a)(-b)}$

<u>Working out the solution:</u> Just like before, we'll start by tackling just the first term. Counting, we find three negatives, so the product of the first term will end up having a negative sign: $(-1)(-a)(-b) = -ab$. Next, we'll look at the second term all by itself: $\dfrac{-a}{(a)(-b)}$. Counting, we get two negatives, which means that when simplified, this term won't end up with a negative sign, so we can drop them. We can proceed by canceling the common factor of a on the top and bottom of the fraction: $\dfrac{-a}{(a)(-b)} = \dfrac{a}{(a)(b)} = \dfrac{\cancel{a}}{\cancel{(a)}(b)} = \dfrac{1}{b}$. Now we can put it all together.

<u>Answer:</u> $(-1)(-a)(-b) + \dfrac{-a}{(a)(-b)} = -ab + \dfrac{1}{b}$

2. $(-2)(-x)(y) + \dfrac{yz}{y}$

3. $\dfrac{-10(-a)}{(-5)ab}$

4. $\dfrac{-9c(-d)}{3d} \div \dfrac{c}{(-2)}$ (Hint: See the example on p. 117).

(Answers on p. 320)

When multiplying/dividing with variables, <u>deal with negative signs first</u>: *Count* the number of negative signs and determine ahead of time if each term will have a negative sign left over. Then forget about the signs and finish simplifying.

 Variables work the same way in fractions as regular numbers do. If a variable is a common factor on the top and bottom of a fraction, then you can cancel it just like you would any other common factor.

Zero can never be *on the bottom of a fraction*, so if x is in the denominator, $x \neq 0$; otherwise, the expression is "undefined" in the world of math. It doesn't *have* a value!

Chapter 9

Do You *Like Him* Like Him?

Combining Like Terms

When I was 14, during my first TV show, *The Wonder Years*, we did an episode centered around a tangled mess of trying to figure out who had crushes on different people. Now, we could have just said, "Do you like him?" But as you know, that question can be ambiguous. So, throughout the whole episode, whenever anyone said they *liked* someone, we'd say things like, "Yeah, but do you like him, or do you *like him* like him?" And yes, this has something to do with pre-algebra.

You might not have known this, but two math terms *like* like each other when they have the same variable part. Yeah they get full-on crushes—it's really cute. For example, $3x$ and $2x$ *like* like each other. The same goes for $4xy$ and $-7xy$. Their coefficients don't matter, just their variable parts. That's how they determine whether or not they're a good match. Notice that $3a$ and $4ab$ are *not* a good match.

When terms have the same exact variable part—when they *like* like each other—then they are considered to be **like terms**. And when we have two or more **like terms** in an expression, we can combine them, which simplifies things quite nicely. The whole expression becomes shorter and neater.

I don't know about you, but I'm a big fan of "shorter and neater" when it comes to math!

"Ring Ring" What's It Called?

Like Terms

Like terms are terms that have the *same exact variable part* as each other.

For example: $8xy$ and $-\frac{1}{2}xy$ and $(0.03)xy$ are like terms, because they all have the same exact variable part: xy.

However, $8xy$ and $8x$ are *not* like terms, because their variable parts are different. Also, $2x$ and $2x^2$ are *not* like terms, because their variable parts are not exactly the same. Exponents do matter!

When we have **like terms**—that is, terms that have the same exact variable parts—we can combine them by combining their coefficients, just like we learned in the coefficient shortcut on p. 101. For example, $5xy - 2xy = 3xy$.

Watch Out!

It might not always be so easy to tell which terms are the like terms. For example, $3xy^2$ and $4y^2x$ are like terms, because xy^2 is actually the same thing as y^2x. Remember, the order of multiplication doesn't matter. What matters is that they both have x and y^2 multiplied together. On the other hand, $3xy^2$ and $4yx^2$ are *not* like terms, because no matter how you rearrange those variables, they'll never be a match. (Some relationships just weren't meant to be.) This is part of why it's a good idea to write your variables in alphabetical order; it makes it easier to keep things straight.

More Bags of Pearls

If you're ever tempted to combine *un*like terms, read this.

When you're simplifying an expression like $3x + 2x + 6$, you probably know that you can combine the first two terms, $3x + 2x = 5x$, so that the full expression would become $5x + 6$, right? That's because x just stands in for a bag of pearls, an unknown amount, and you know that 3 bags + 2 bags = 5 bags.

However, if you have $3x + 2y + 6$, there's nothing you can do to simplify it further, because x and y are two *entirely different kinds of bags* of pearls. In fact, let's call y a "sack" of pearls. The "x bag" is holding some unknown number of pearls, say 5 or 10, right? Meanwhile, the "y sack" might be holding $2\frac{1}{2}$ pearls, or a billion pearls for all we know. In other words, because x and y are different variables with *different* unknown values, we can't combine them to simplify the expression any further.*

Do You Doodle?

If you were asked to simplify something like this by combining like terms:

$$2x + y + 2y$$

...you could probably just go ahead and write the answer: $2x + 3y$. You'd be like, "Hey, this combining like terms stuff isn't so bad!"

However, before you know it, your teacher will give you long expressions like this, and say, "Simplify":

$$3x + y - x^2 + 2 - 7 - 4y + 5x^2 + x + 2y$$

And you'll be like, "Yeah, sure. Easy for you to say." Well, there's a good way to handle this type of thing, and believe it or not, it actually involves a little doodling.

Before we explore our artistic side, however, we need to rewrite ALL the subtraction as "adding negatives." We'll be so glad we did. And this way, we can move things around freely and get out of any PEMDAS confusion.[†] Let's also write in the sneaky 1 and −1 coefficients:

$$3x + 1y + (-1x^2) + 2 + (-7) + (-4y) + 5x^2 + 1x + 2y$$

Now for the doodling. Your job is to go through this and look for *like terms*. Focus only on the *variable parts* of the terms, and find the ones that are exactly the same as each other. Let's start with the x terms. Now we get to choose a creative style of underlining for them. How about this? So each time we see an x term, we'll give it that style of underlining.

Next variable? How about the y terms? Let's give them this kind of underlining, and how about this for the x^2 terms? This is the best way I've found for keeping track of like terms! You don't have to keep too much in

. . . .

* By the way, xy would be yet another, totally different kind of bag—perhaps a purse or a satchel—and would hold an entirely different number of pearls in it, too.
† Check out the WHAT'S THE DEAL? on p. 35 for more on this.

your head at once. Plus, it's kind of like doodling, which has always been one of my favorite hobbies.

$$3x + 1y + (-1x^2) + 2 + (-7) + (-4y) + 5x^2 + 1x + 2y$$

And now we're ready to combine terms. First, for the *x* terms, we look for the "crosses" underlines and combine $3x + 1x =$ **4x**. Now for the *y* terms, we look for the wavy underlines: $1y + (-4y) + 2y =$ **−1y**. And we know that $-1y = $ **−y**. Next, for the x^2 terms, look for the zigzag underlines: $(-1x^2) + 5x^2 =$ **4x²**. Now go through and make sure we got 'em all. Yep!

Finally, we can do the constant terms. You can either underline them or not; they stand out on their own pretty well. It's your choice. Here, we get $2 + (-7) =$ **−5**. Putting it all together, we get **4x + (−y) + 4x² + (−5)**. And that's the answer. Phew!

Watch Out!

Notice that x^2 is also its own kind of "bag of pearls," so: $x^2 + x^2 = 2x^2$. Don't be tempted to say it equals x^4! (You'll learn more about variables with exponents in Chapter 16.)

QUICK NOTE   In the previous example, we also could have written the "adding negatives" back into subtraction if we had wanted to (or if that's what the teacher wants) so the answer would look like this: $4x - y + 4x^2 - 5$. But if you ask me, it's fine to leave them as "adding negatives" until you get into algebra later on.

Watch Out!

Make sure you use very distinctive underline styles so they look different from each other and you don't get confused! You can always use <u>single</u>, <u>double</u>, and <u>triple</u> underlines if you know your handwriting won't be neat enough or if you're not feeling very artsy.

Step By Step

Simplifying expressions by combining like terms:

Step 1. First, rewrite all subtraction as "adding negatives." (See p. 7 for a review!) If it helps, write in 1 and −1 where coefficients aren't written in front of variables.

Step 2. Give each type of **like term** its own underline style.

Step 3. Combine **like terms** by combining their coefficients. (See p. 101 for a review.) Done!

Step By Step In Action

Let's simplify this expression by combining like terms:

$$(0.5)ab^2 + 2b^2 + (1.5)ab^2 + 8 - ab^2 + b^2 + 2 - 3b^2$$

Steps 1 and 2. Yikes! Okay, let's first rewrite the subtraction as "adding negatives," and then write in the 1 and −1 coefficients. Then we'll <u>underline</u> like terms. Pay attention and you'll notice that there are **two** types of variable terms: ab^2 and b^2.

$$(0.5)ab^2 + 2b^2 + (1.5)ab^2 + 8 + (-1ab^2) + 1b^2 + 2 + (-3b^2)$$

Step 3. Now we're ready to combine like terms. Let's start with ab^2 terms: $(0.5)ab^2 + (1.5)ab^2 + (-1ab^2)$. Remember, we can just think of this as combining their coefficients: $0.5 + 1.5 + (-1) = 2 + (-1) = 1$. So that means $(0.5)ab^2 + (1.5)ab^2 + (-1ab^2) = 1ab^2$, which is the same as $\boldsymbol{ab^2}$, right?

Okay, next we'll combine the b^2 terms:

$$2b^2 + 1b^2 + (-3b^2) = 0b^2 = \mathbf{0}$$

Hey, how did that happen? Remember, zero times anything equals zero! Or you can think of it like this: 2 bags plus 1 bag, minus 3 bags, leaves us with no bags. Anyway, our b^2 term just went away. So that's nice. Next, let's combine our constants: $8 + 2 = \mathbf{10}$. So we get $ab^2 + 0 + 10 \rightarrow \boldsymbol{ab^2} + \mathbf{10}$. Ah, much nicer!

Answer:
$$(0.5)ab^2 + 2b^2 + (1.5)ab^2 + 8 - ab^2 + b^2 + 2 - 3b^2 = \boldsymbol{ab^2} + \mathbf{10}$$

Doing the Math

Simplify these expressions by combining like terms. I'll do the first one for you!

1. $\frac{1}{5}x - 4x^2 - (0.2)x + 3x^2$

<u>Working out the solution</u>: First, we'll rewrite the expression by changing subtraction to "adding negatives," and then we'll underline *like terms*: $\frac{1}{5}x + \underline{(-4x^2)} + \underline{(-0.2x)} + \underline{3x^2}$. Now we can combine like terms, so let's do the x terms first: $\frac{1}{5}x + (-0.2x)$. We just combine the coefficients, so that's $\frac{1}{5} + (-0.2)$. In order to combine these, we need to convert them both to fractions or decimals, right?* Let's change -0.2 to a fraction by first rewriting it like this $-\frac{0.2}{1}$ and then multiplying it times the copycat fraction† $\frac{10}{10}$ so we make sure not to change its value:

$$-0.2 = \frac{-0.2}{1} \times \frac{\mathbf{10}}{\mathbf{10}} = \frac{-0.2 \times \mathbf{10}}{1 \times \mathbf{10}} = \frac{-2}{10} \text{ (reducing)} = -\frac{1}{5}$$

Okay, now we can combine x's coefficients: $\frac{1}{5} + \left(\frac{-1}{5}\right) = 0$. So that means $\frac{1}{5}x + (-0.2x) = 0x = 0$. So much for the x term, huh? It's so nice when that happens. Next, let's do the x^2 terms, which will be much simpler: $(-4x^2) + 3x^2$. Combining coefficients, we get $-4 + 3 = -1$.

So $(-4x^2) + 3x^2 \rightarrow -1x^2 = -x^2$.

<u>Answer</u>: $\frac{1}{5}x - 4x^2 - (0.2)x + 3x^2 = -x^2$

. . . .

* To review converting decimals to fractions, see Chapter 12 in *Math Doesn't Suck*.
† I like to call fractions with the same top and bottom "copycat fractions." Copycat fractions always have the value of 1. For more on this, see p. 64 in *Math Doesn't Suck*.

2. $5 - g + 2h + 2g - h$

3. $6a + 7b + b^2 - 2a + 3b - 7b^2$

4. $3x + 3yz - 3xy - 3x - 3zy$ *(Hint: Be careful!)*

(Answers on p. 320)

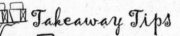

Takeaway Tips

 Like terms are terms whose variable parts are exactly the same, exponents and all. Watch out for when the variable parts aren't written in alphabetical order. Sometimes they can be deceptive. For example, $4x^2y$ and $9yx^2$ are actually like terms!

 Like terms can be combined either by thinking about how they represent bags of pearls or by just combining their coefficients.

When you have a long string of terms to combine, go through and use a different kind of underlining for each type of *like term*, which will make it much easier to combine them and keep them straight.

TESTIMONIAL

*Maria Quiban** (Los Angeles, CA)

Before: Distracted by friends
Today: Fabulous TV personality and meteorologist

Until the ninth grade, I loved math. It was fun and exciting—exhilarating, even. Algebra particularly interested me; it was like learning a new language. I loved the way math sharpened my mind and the confidence this gave me. But then math got harder, and I started hanging out with my friends more. I got really distracted by my social life, and I wanted to seem "cool." I even missed classes sometimes, so of course I was constantly struggling in math.

> "I'd say to myself, 'If they can do it, so can I!'"

In college I realized I wanted to become a broadcast meteorologist (you know, the man or woman who reports the weather on TV in front of colorful animated weather graphics), and I knew that required lots of math. This time, I never missed a class, and I took full advantage of group study and tutoring. Being "cool" wasn't cool anymore. Instead, I kept my eyes on the prize—my dream of becoming a meteorologist—and before long, I was able to solve and understand equations that I wouldn't have even attempted in the past. And I'm glad I did!

Today, I'm a successful broadcast meteorologist at a top television station in Los Angeles—and I love it! My broadcasts every night are live, and the adrenaline rush you get when you're covering breaking weather news is an invigorating experience. I love explaining weather phenomena as it's

* Photo courtesy of Fox Television Stations, Inc.

happening—helping viewers understand the current conditions and warning them of areas where weather might be a danger. My job allows me to express my artistic side, too; I use the latest computer software in weather animations to create graphics and produce my segments.

I take great pride in gathering the very latest weather data and creating the best forecast possible for the TV viewers, and it's in creating my weather forecasts that math really comes in handy. Weather data—temperatures, winds, storms—it's all numbers, and it's my job to interpret those numbers!

I use math every day to quickly convert knots to miles per hour, Celsius to Fahrenheit, and more. My job involves not only the substitution of variables into simple formulas, but also interpreting numerical models that are generated by high-speed computers using very complex equations. In weather reporting, there's a lot of math that goes on behind the scenes!

I'm proud to be Filipino American, one of the many faces representing and promoting diversity in Los Angeles TV, and math was an essential ingredient for this dream to come true. My success in college math didn't come easy at first. It came from my willingness and commitment to embrace the material 100%—to not let anything intimidate me. When I needed motivation, I would simply look at the other kids in my class, and I'd say to myself, "If they can do it, so can I!"

And you know what? So can *you*.

Chapter 10

The Costume Party
The Distributive Property

By the end of this chapter, you'll be a whiz at the distributive property and how it can help with simplifying all sorts of expressions. It's a tool I use all the time; it's even kept me from getting ripped off! I'll tell you a true story about that at the end of this chapter.

You might already know the word *distribute*. Like, if you have to pass out homework assignments to everyone in your class, your teacher might ask you to *distribute* them to your classmates. (How'd you end up with that job?)

Okay, now imagine you're going to a costume party at your friend Beth's house. She's dressed up as a bride, so we'll call her *b*. You're almost the first to arrive; there's only one other guest, but you can't tell who she is because she's dressed up as a cat with a full-on cat mask. So their house looks like this: $(b + c)$, where the *b* stands for bride, the *c* stands for cat, and the parentheses are the walls of the house. Just go with me on this one!

You're dressed up as Ariel from *The Little Mermaid*, so we'll call you *a*. So when you knock on their door, just outside the house, it looks like this: $a(b + c)$, right? Once you get inside, naturally you say hello and give a hug to each friend!

You, saying "hi"!

$$a(b + c) = ab + ac$$

And if there'd been another guest there, perhaps one dressed up as a dragon, well you'd say hi to her, too! After all, you're very polite.

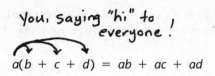

You, saying "hi" to everyone!

$$a(b + c + d) = ab + ac + ad$$

Maybe you didn't know this, but numbers and variables are also very polite. They always say hi to everyone at a party, too.

In math, when you see $a(b + c)$, this means that the a is multiplying* times the $(b + c)$. In homework or on tests, you might be asked to "distribute the a over the parentheses" or more specifically here, "distribute the a <u>over</u> b and c." This means they want you to use the distributive property (see WHAT'S IT CALLED? below) to *rewrite* the expression so that the a is directly multiplying times the b and also times the c. This is just like when you say hi to your friends!

By the way, inside the parentheses, you can have either addition or subtraction. The distributive property works the same for both.

"Ring Ring" What's It Called?

Distributive Property (Of Multiplication over Addition and Subtraction)

The *distributive property* is a rule that allows us to *rewrite* very specific kinds of expressions (like the ones below) so that the parentheses go away but the value of the expression remains unchanged. Think of coming to a party and saying hi to each friend!

For any numbers a, b, and c, here's how it works:

With addition: $a(b + c) = ab + ac$
With subtraction: $a(b - c) = ab - ac$

. . . .

* Multiplication can look a few different ways, so $a(b + c)$ is the same thing as $a \times (b + c)$. See the box on p. 108 for a refresher. Don't worry; this is confusing at first!

By the way, *a*, *b*, and *c* are just acting as placeholders. They could be standing in for numbers or even more complicated terms like $5x^2$. But we'll get to that soon!

Watch Out!

If you have $a + (b + c)$, there's nothing to distribute. You only distribute when the number on the outside is *multiplying* times the parentheses. So the *a* has to be touching it, ringing the doorbell so to speak, to get into the party!

Not all parties are costume parties, but some people like to dress up anyway. You might be the 2 coming to the party like this: $2(3 + x)$. Let's see, there's your friend 3, and . . . um, looks like someone's still wearing a costume. Of course you say hi to her anyway, even though she's wearing a bunny costume to a pool party. You're polite, remember?

Let's use the distributive property now to rewrite the expression $2(3 + x)$, distributing the 2 to each term inside. Try to imagine the little arching "saying hi" arrows.

$$2(3 + x) = 2(3) + 2x = \textbf{6} + \textbf{2x}$$

What about something like $x(y - 2)$? We could change the subtraction to "adding a negative" (see p. 7) and then distribute the *x* over addition to each term inside:

$$x(y - 2) = x(y + [-2])$$
$$= xy + x[-2] = \textbf{xy} + \textbf{(-2x)}$$

Or, just following the distributive rule for subtraction (p. 131), we could simply distribute over the subtraction:

$$x(y - 2) = xy - x(2) = \textbf{xy} - \textbf{2x}$$

Both answers are the same!

Forgetting to include negative signs is one of the most common mistakes in using the distributive property, so keep an eye on 'em.

QUICK NOTE A word about brackets vs. parentheses: Brackets [ ] and parentheses ( ) are interchangeable; you can use either one. So, 7[x + 4] means the exact same thing as 7(x + 4). Just like in the prior example, when there are more complicated groupings, it's really helpful to use both to keep track of what's going on! For example, 2[x + (−0.5)] is a bit easier to decipher than 2(x + (−0.5)), right?*

Believe it or not, the distributive property can be helpful in life. (See Danica's Diary on p. 142.) *Sometimes* it can even make fractions and decimals easier.

QUICK NOTE Just like the associative and commutative properties we learned in Chapter 2, notice that the distributive property is another way of breaking the PEMDAS order of operations. (See p. 21 for a review of the Panda PEMDAS rules.)

Doing the Math

Use the distributive property to break the PEMDAS rule. In most other problems, you'd want to simplify inside the parentheses first. But in these *particular* cases, I want you to **distribute first**, to get practice. And distributing first will make some of these problems much easier. I'll do the first one for you.

* * * * * * * * * * *

* They both equal $2x - 1$, by the way.

1. $9\left(\frac{2}{3} - \frac{5}{9}\right) = ?$

<u>Working out the solution</u>: We'll distribute the 9 over both terms inside the parentheses and then simplify:

$$\frac{9}{1}\left(\frac{2}{3}\right) - \frac{9}{1}\left(\frac{5}{9}\right) = \frac{9 \times 2}{1 \times 3} - \frac{9 \times 5}{1 \times 9}$$

(cancel and reduce)

$$= 6 - 5 = 1$$

<u>Answer</u>: $9\left(\frac{2}{3} - \frac{5}{9}\right) = 1$

2. $14\left(\frac{8}{7} + \frac{3}{14}\right) = ?$

3. $10(8.1 - 4.9) = ?$ (*Hint: remember what happens to the decimal point when you multiply by 10.*)

4. $10\left(8.1 - \frac{1}{5}\right) = ?$ (*Hint: If you do this correctly, it should be pretty easy.*)

(Answers on p. 320)

More About the Distributive Property

At parties, some people are attached at the hip. Have you ever noticed that? They're like a single unit. When you say hi to them, you can pretty much kill two birds with one stone and just say hi once. They even hug you together. So that cuts down the time factor when you're first saying hi to everyone at a party.

On p. 90, we talked about *terms*, that is, expressions stuck together with only multiplication or division. These terms are like groups of friends that act like a single unit. In multiplication, the numbers and variables almost look like they're kissing, don't you think? Look at (8)(4) and 4x or even $5x^2$. And in division, like with $\frac{9}{5}$ or even $\frac{x}{7}$, it looks like someone's riding piggyback, so those numbers seem pretty "close," too. Everything's all stuck together, so it's a single unit—a single *term*.

The point of all this is that when you distribute over a complicated-looking expression, make sure to distribute only *once* to each *term*. For example:

$$3\left[\frac{x}{7} + (2)(5)\right] = 3\left(\frac{x}{7}\right) + 3(2)(5) = \frac{3x}{7} + 30$$

Notice how we distributed the 3 only once to each term. A mistake would be to distribute the 3 to the 2 and *separately* to the 5. See what I mean? We'd get the wrong answer!

Also, sometimes the term on the <u>outside</u> of the parentheses can be sort of complicated (a *very* close group of friends arrives at the party; they do *everything* together). Here's an example of that:

$$3xy\left[\frac{x}{7} + (2)(5)\right] = 3xy\left(\frac{x}{7}\right) + 3xy(2)(5) = \frac{3x^2y}{7} + 30xy$$

In the following example, the only operations inside the parentheses are multiplication and division,* so the expression is still just *one* term here, a very friendly single unit you can say hello to all at once. In other words, you really don't "distribute" the 2 at all:

$$2\left[\frac{x \cdot 5w}{y}\right]$$
$$= \frac{2}{1}\left[\frac{x \cdot 5w}{y}\right] = \frac{2 \cdot x \cdot 5w}{1 \cdot y} = \frac{10wx}{y}$$

Just squish it all together as you normally would for multiplication. Remember, distribution happens across addition and subtraction, *not* multiplication or division.

You might not see some of these more complicated expressions until algebra, but there are some tough teachers out there, so this way you'll be prepared. Just keep thinking about the party, and you'll be fine!

. . . .

* In fraction form.

Distributing Sneaky Negatives

Negatives can be really sneaky. When you see something like $-(b - 3)$, that negative sign is actually -1 that's hiding and that needs to be distributed to the terms inside the parentheses, just like you would for *any other number* stuck to the outside of the parentheses.

$$-(b - 3) = (-1)(b - 3) = (-1)(b) - (-1)(3) = -\boldsymbol{b} + 3$$

You can also first change the inside subtraction to "adding a negative" and then distribute the -1 to the inside terms, over addition instead of subtraction:

$$-(b - 3) = (-1)(b - 3) = (-1)[b + (-3)] = (-1)(b) + (-1)(-3)$$
$$= -\boldsymbol{b} + 3$$

Both methods give you the same answer, so use whichever makes more sense to you. Notice how the signs of the terms inside all change from positive → negative, and negative → positive. Once you get more practice, you can think of distributing a negative as simply switching each of the inside terms' signs: $-(b - 3) = -b + 3$.

Sometimes these sneaky negative signs even disguise themselves as subtraction; in fact, this happens all the time. Hey, who invited them to the party, anyway?

Even Sneakier Negatives: Disguised as Subtraction

When you see something like this:

$$9 - (x + 3y)$$

. . . that subtraction is actually a negative sign that *needs to get distributed* inside the parentheses! I highly recommend rewriting subtraction as "adding negatives"* in order to keep everything straight. Check it out.

If we had $9 - (15)$, we could rewrite it as $9 + (-15)$, or even $9 + (-1)(15)$, right? (Take the time to "get" that before reading on.) We'll do the same thing with $9 - (x + 3y)$. Pay attention to the negative signs and the parentheses. If you get confused during the first step, just imagine $(x + 3y)$ replaced by (15), and it should make sense why we can do this:

$$9 - (x + 3y)$$
$$= 9 + (-1)(x + 3y)$$

(Now we can distribute the (-1) to each term inside the parentheses.)

$$= 9 + (-1)(x) + (-1)(3y)$$

(Now we just complete the multiplication.)

$$= \mathbf{9 + (-x) + (-3y)}$$

(We could also write the answer like this; notice that they are equal!)

$$= \mathbf{9 - x - 3y}$$

And voilà! It looks like we out-sneaked the negatives after all.

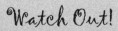

Watch Out!

A common mistake is to distribute $-2(x - 3)$ and get $-2x - 6$. But this is wrong! Do you see why? The negatives on -2 and -3 should cancel to give us positive 6. Remember: If there's subtraction inside the parentheses, you can always rewrite it as "adding a negative" if that makes things clearer to you.

$$-2(x - 3) = -2(x + [-3]) = -2x + (-2)[-3] = \mathbf{-2x + 6}$$

. . . .

* See p. 7 for a review.

Using the distributive property correctly:

Step 1. Simplify the inside of the parentheses if needed. (Get rid of any easy messes to clean; it will save you time in later steps!) Also, rewrite subtraction as "adding negatives" anywhere it will help you.

Step 2. Notice whether or not the term on the outside of the parentheses has a *negative sign*. If it does, make a mental note that the *signs* of all the terms *inside* the parentheses will be changing.

Step 3. Distribute the multiplication to *each term* inside the parentheses, keeping an eye on the negative signs. Done!

 *And...* *Action!* *Step By Step In Action*

Let's use the distributive property to rewrite this expression:

$$-\frac{1}{5}(5a - 10 + b)$$

Step 1. There's nothing to simplify inside, and the only subtraction happens before the 10, so let's rewrite that as adding a negative: $-\frac{1}{5}(5a + [-10] + b)$. So far, so good!

Step 2. Because $-\frac{1}{5}$ is negative, we know the *signs* of the inside terms will be changing.

Step 3. Distribute $-\frac{1}{5}$ to all *three* terms inside the parentheses, watching for negatives. (I added parentheses around each $-\frac{1}{5}$ so that we didn't end up with "+ −" anywhere.)

$$-\frac{1}{5}(5a + [-10] + b) =$$

$$\left(-\frac{1}{5}\right)(5a) + \left(-\frac{1}{5}\right)[-10] + \left(-\frac{1}{5}\right)(b) =$$

$$-a + 2 - \frac{b}{5}$$

And voilà! These negatives can be sneaky, but once we disarm them by switching subtraction to "adding negatives," *we* get the upper hand, oh yes we do.

Take Two: Another Example

Let's simplify something sort of scary looking, like this:

$$11 - [-3 - (-9x) + x]$$

Step 1. Let's simplify inside the brackets. I mean, just *look* at all those negative signs. Something must be done about them!

$$-3 - (-9x) + x \rightarrow -3 + 9x + x \rightarrow \mathbf{-3 + 10x}$$

Now let's rewrite our problem with the new, simplified expression inside the brackets and continue with our step-by-step method:

$$11 - [-3 + 10x]$$

Step 2. Yep, there's a negative sign outside the parentheses, and we know it's really a sneaky (-1). Let's write it like that:

$$11 + (-1)[-3 + 10x]$$

Step 3. Now let's distribute that negative sign into the parentheses, and we'll get

$$11 + (-1)(-3) + (-1)(10x) = 11 + 3 + (-10x)$$
$$= \mathbf{14 + (-10x)}$$

Or, $14 - 10x$. Done!

Sometimes this stuff can get sort of confusing looking because of all the variables and negative signs and parentheses. But if you take it one step at a time and remain *calm*, you'll build some pretty strong muscles that'll make algebra a breeze.

Doing the Math

Use the distributive property to simplify these expressions. And watch those negative signs. I'll do the first one for you.

1. $-3xy - 3x(2 - y)$

<u>Working out the solution</u>: Hmm, lots of negative signs. No problem: We'll change the first subtraction into "adding a negative" and get a handle on things: $-3xy + (-3x)(2 - y)$. Let's distribute the $-3x$ inside the parentheses, and we'll get $-6x + 3xy$. Now our full expression is $-3xy - 6x + 3xy$, right? Combining *like terms* for $-\underline{3xy} - 6x + \underline{3xy}$, we end up with $-6x$. And that's about as simplified as it gets!

<u>Answer</u>: $-3xy - 3x(2 - y) = -6x$

2. $5 - (h - 4)$

3. $10 - 3y(x - 4)$

4. $xy - 10\left(0.8 + \dfrac{xy}{10}\right)$

5. $8ab - a\left(b - \dfrac{1}{a} + 3\right)$

(Answers on p. 320)

What's the Deal?

Why are we *allowed* to distribute multiplication over addition and subtraction? And how can we be sure that $a(b + c)$ has the *same exact value* as $ab + ac$?

Let's apply this so-called distributive property to a really easy number problem and see what happens: $3(10 + 1) = ?$

$$3(10 + 1) = 3(10) + 3(1) = 30 + 3 = \mathbf{33}$$

And that made total sense, right? Because after all $3(10 + 1) = 3(11) = 33.$

Now let's do one for an easy subtraction problem: $4(10 - 1) = ?$

$$4(10 - 1) = 4(10) - 4(1) = 40 - 4 = \mathbf{36}$$

And lookie there: $4(10 - 1) = 4(9) = 36$ is also pretty familiar. This little demonstration shows us that, yep, the distributive property really works. You can always do one of these easy problems to remind yourself of what the rule is or how it works!

Takeaway Tips

The distributive property is like someone going to a party and saying hello to everyone (each term) inside!

Remember that a *term* is anything that's stuck together with multiplication or division. Terms are separated from each other by addition or subtraction.

When using the distributive property, be especially careful when dealing with subtraction and those sneaky negative signs: To be safe, you can always rewrite subtraction as "adding negatives" and also remember that a negative sign stuck to the outside of a parentheses is the same as (-1) multiplying the parentheses.

Danica's Diary

ALMOST GOT RIPPED OFF!

This is a totally true story. A few months ago, I was shopping at a department store, and I found the cutest top. I wanted to get it in three colors for myself, plus one for my sister and one each for two friends. Hey, birthdays were coming up, and a gal has to be prepared!

So, I'm standing there with my six tops, waiting in the checkout line and feeling bored. I thought I'd go ahead and figure out how much my bill would be. After all, I'm always encouraging people to do math in their head, so I thought maybe I should take my own advice!

Each shirt, before tax, was \$38. But I wasn't in the mood to multiply 6×38, and I *really* wasn't in the mood to add up $38 + 38 + 38 + 38 + 38 + 38$. So guess what I did? I thought about how 6×38 is like $6(40 - 2)$, and I used the distributive property to do the math in my head.

I knew that $6 \times 4 = 24$, which meant that $6 \times 40 = 240$. So far, so good. Now, I knew I still needed to subtract 6×2 from it. So, I needed to figure out $240 - 12$, which isn't too hard to do in your head when you focus: $240 - 12 = \$228$. Alright. That would be the amount *before* tax.

I reminded myself that math in my head is like doing push-ups for my brain, and I kept going. Plus, there were still three people ahead of me in line.

With California tax at about 8%, which is 0.08, I would need to multiply 0.08 times \$228. I knew that figuring out 10% of \$228 would just involve moving the decimal place once: \$22.80. I'll be honest with you, at this point I allowed myself to round it to \$23. And I knew that 1% would just require moving the decimal place over twice: \$2.30. That means that 2% should be twice that, which is about \$4.60. I rounded again to \$5. I still hadn't gotten up to the front of the line yet, so I figured out that 8% would be 10% − 2% = \$23 − \$5 = \$18. So that would be the *tax* on my items. What was the total again? Oh, yeah: $228. Adding approximately $18 of tax, I got $246 or so.

Well, finally, I'm up at the front. And my brain is hurting, but I'm glad I toughed it out. Push-ups are good for me, right? I'm all distracted, feeling good about my push-ups as I hand the guy my credit card. I'm about to sign the receipt, and I see—to my utter disbelief—that he's rung me up for $287.28! I look up and try to be as polite as I can: "Excuse me, I think this amount might be wrong." The guy was like, "No, it's fine. The computer does it." I said, "Um, okay, but maybe you rang up one of the tops twice? This seems like more than it should be." He looked at me like I was five years old and said, "There's tax added, you know."

I kid you not. This guy totally fought me on it. But I had just figured it out, and I knew I was right! Either that, or I'd made some mistake in my head, which was possible, but I wanted to see for myself. What good is math if it can't save you from getting ripped off? I politely asked for the itemized receipt and saw that he had indeed scanned in seven tops, not six. He'd double-scanned one of them!

And I never would have known that $287.28 should have been closer to $246 if I hadn't done that math in my head. And thanks to the distributive property, I did.

After I showed him the mistake on the receipt, he turned it over to his manager, because he didn't know how to cancel a credit card transaction. Maybe that's why he was so stubborn about it. Remember, cashiers just scan in items and the register does the rest. They're not paid to check the math. It makes you realize how often things are probably rung up incorrectly and no one ever catches it!

Every time you go shopping, whether it's for clothes or whatever, try adding up the amounts in your head while you're standing in line. Do it with approximations if you need to, but get in the habit of doing math in your head. I want to hear about the first time it saves you from getting ripped off! Just email me at headmath@kissmymath.com.

Fitness Tips
That Take ZERO Time

Check out these zero-time fitness tips from Chicago-based expert physical fitness trainer Lori Verta. Her tips are great for strengthening your body and fighting stress. Just like doing math in your head, you can do these while standing in line or brushing your teeth, and they take *zero* time out of your day!

Wing Squeeze: Imagine your arms are wings that start in the middle of your back, and squeeze them gently together behind you. Hold for a count of 2. Repeat this five times, or whatever is comfortable. Be sure to press your shoulders down, so they don't move up. Also, breathe very deeply and feel your chest expand as you do. This exercise should feel good! We all spend so much time at our computers and cradling cell phones, and this much-needed exercise strengthens the muscles around our shoulder blades, helping relieve built-up stress. It also helps correct our posture. When we are stressed, the first thing to go is our posture. We tense up, and become unaware of how we are sitting or standing or what we are doing with our body position. Who wants to walk around all hunched over? Not me!

Glute Squeeze: Simple—just tighten and relax your rear! This one can be done just about anywhere, and no one will know you're doing it! Be sure to keep breathing during the exercise. Strengthening our gluteal muscles creates a stronger base for our upper body and helps correct our posture as well!

Toothbrush Balance: While you brush your teeth, try standing on one leg. Hold onto the counter for support, and notice how your leg muscles have to work harder to keep you steady. For more challenge, try letting go of the counter and notice how much harder your leg is now working! But never endanger yourself; always use caution and be safe with this one. And always, always make sure the bathroom floor is dry so you don't slip. This is a simple, effective exercise for leg strength and balance. And the best part is—just like doing math in your head while you're standing in line—it makes you *stronger*, but it doesn't take up any extra time!

Chapter 11

Didn't That Guy Say He Was Going to Call?

Using Variables to Translate Word Problems

Math is its own language, and we must become expert translators—from English to math, and math to English—in order to solve the word problems (aka, story problems) that'll come up from now through your years in algebra class and beyond. In this chapter, we're going to get good at *translating*, which can be the most challenging and important part of word problems! Actually, translating comes in handy for more than just word problems.

Girls and guys *seem* to speak the same language, and yet, somehow, I've always had the feeling that we don't. For example, when a guy says, "I'll call you," it could very well mean, "I'm not going to call you." Very confusing language they use; it can be difficult to translate guy-speak. This is why so many girls rack up cell phone bills trying to dissect what a guy has just said and what it means. If it were straightforward, then why would it take hours to analyze? I think you're seeing my point now.

To be fair, girl-speak probably isn't very clear, either. If we say to a guy, "You don't need to get me anything for my birthday," it really means, "You need to get me something for my birthday. And now you've been officially reminded that my birthday is coming up." If he so much as blinks the wrong way, the silent message continues: "It doesn't need to be expensive, but I mean, you're going to get me *something*, right? Unless of course you're saying you want to break up. Do you want to break up?"

See what I mean?

Luckily, translating math-speak is a bit more straightforward, so let's at least get good at *that* for now. More on girl-guy stuff later in the chapter.

Translating Math → English

Math Sentences

Something important to realize about the math language is that, just like in English, there can be full sentences, and there can also just be fragments. *Like this.* That wasn't a sentence—it was just a fragment—and it didn't say much on its own, did it?

Full math "sentences" are things like equations and inequalities. They *say* something. They compare two things and might reveal the information: "This <u>equals</u> that" or "This <u>is</u> bigger than that." I underlined the verbs so you can see them.

See? Yep, they're sentences, all right.

Math Sentences

What Are They Called?

Equation

An *equation* uses an equals sign (=) and is a statement of truth and equality. That sounds rather noble, doesn't it? An equation has something to *say*. It's a math sentence; it says, "This <u>is</u> equal to that." It gives us information.

Here's an example of an equation: $10 = 8 + 2$. Yep, the thing on the left equals the thing on the right. Here's another example of an equation: $x + 2 = 5$. Again, this equation tells us that the thing on the left *is equal* to the thing on the right (which also tells us that x's value must be 3). More examples of equations:

$$y = 9 \text{ means } "y \text{ } \underline{is} \text{ equal to nine."}$$

$$7x - 13 = x + 17 \text{ means}$$
"Seven times x, minus thirteen <u>is</u> equal to x plus seventeen."

Inequality

An *inequality* is also a statement of truth—a math sentence. Some inequalities say "This thing over here <u>is</u> bigger than (or less than) that thing over there" and use the symbols > or <. For example, $x < 9$ translates to "x <u>is</u> less than 9."

Being math *sentences*, inequalities also give us information; they *say* something. I mean, we might not know what x's exact value is, but we do know that it <u>is</u> less than 9. That's information, right? Other inequalities include the option of the two sides being equal, and those would use the symbols $\leq$ or $\geq$. Here are some examples:

$2 < 3$ means "Two <u>is</u> less than three."

$x > 0$ means "x <u>is</u> positive."

$2y \leq 8$ means "Twice y <u>is</u> less than or equal to eight."

If those are math sentences, then what's *not* a math sentence? Glad you asked.

"Ring Ring" What's It Called?

Expression

An *expression* is not a sentence of any sort, so it *doesn't get the "is" verb*, like the prior translated math sentences do. It's just one or more math terms and operations, sort of hanging out. An expression is just kinda . . . there.

Here are some examples of expressions: $2x + 3$ and $9(x - 8)$. See how they just sit there? Nothing is being compared to anything else; nothing is equal to anything else; they don't give us any information. It's almost like they're sitting there making faces at us! You know, with funny *expressions* on their faces. (That's one way to remember that these are called expressions).

QUICK NOTE   Notice the difference in what we can "do" with variables that are in expressions, as opposed to what we do with variables that are in math *sentences*.

Expressions: We can *substitute values* into the variables in expressions and then *evaluate* them, as on pp. 84–7. But there's nothing to solve for.

Math Sentences (Equations, Inequalities): We solve these; we find the value of x that makes them true. On the other hand, if we tried to substitute values for x, we'd likely get untrue statements. (Just try substituting x = 1 into the equation 2x + 1 = 5. Oops!)

The Math Verbs

In math sentences, the "verbs" look like this: $=, <, >, \leq, \geq, \neq, \approx$. (Pretty funny-looking verbs, if you ask me.) There are others, but these are the ones you'll need in algebra.* Just put two things on either side of one of these "verbs," and you've got a math sentence!

Math Verb	English Translation(s)
$=$	is equal to, equals, is the same as
$\approx$	is approximately equal to†
$\neq$	is not equal to, does not equal
$<$	is less than, is smaller than
$>$	is greater than, is bigger than
$\leq$	is less than or equal to, is smaller than or equal to
$\geq$	is greater than or equal to, is bigger than or equal to

Isn't it amazing how much time and space these math verbs can save us?

His ego $>$ a building.

Notice that you wouldn't say: His ego *is* $>$ a building. That wouldn't be correct, because we already have *is* in the $>$.

I guess the real question is, if you passed this note in math class, using the math verb *correctly*, could the teacher really get mad at you?

Well . . . yes. Especially if the note is about the teacher.

• • • • • • • • • •

* For example, in geometry you'd use "verbs" like these: ∥ (is parallel to), ⊥ (is perpendicular to), ≅ (is congruent to), etc. Saves lots of space!
† This one gets used, for example, when dealing with some unit conversions. We usually say: 1 mile $\approx$ 1.61 kilometers, even though the actual number of digits is much larger: 1.60934 . . . kilometers. Any time we round decimals, we need to use the $\approx$ symbol.

Doing the Math

a. Say whether or not these are math sentences: Are they equations, inequalities, or just expressions? **b.** Then, using the above chart, *translate* these math phrases into English. If there is a math verb, then underline the verb in the English sentence. I'll do the first one for you.

1. $1 < \frac{x}{2}$

<u>Working out the solution</u>: Yes, this is a math sentence; in fact, it's an inequality. It tells us that 1 is less than $\frac{x}{2}$; remember, the alligator mouth always tries to eat the bigger value.

<u>Answer</u>: a. It's an inequality. b. 1 <u>is</u> less than half of x.

2. $2x - 1 = 0$

3. $\frac{y}{3} + 3 + x$

4. $a \geq 2$

5. $g + 0$

6. $\frac{z}{3} < 7$

(Answers on p. 320)

Kind of makes you appreciate the brevity of math, doesn't it? Let's get good at translating from English → math now. It's really a much more efficient language.

First, let's do some warm-ups and exercise the translating muscle. Look for the "is" verb for a clue about whether or not each phrase is a sentence. My other piece of advice here is this: READ CAREFULLY! Every word counts. (Remember, *of* means multiplication. See the chart on p. 152 for more stuff like this, but you shouldn't need it much.)

 ## Doing the Math

a. Underline the "is" verb if there is one, and say whether the phrase is an equation, an inequality, or just an expression.
b. Translate each phrase into math. I'll do the first one for you.

1. Seven less than twice x

<u>Working out the solution</u>: There's no verb here, so that means I can't use $<$, which most people would be tempted to, especially since we're learning about inequalities! But this is not a math *sentence* of any kind; it's just an expression. Reading it carefully, we realize that "less than" must mean subtraction, so how would we express 7 *less* than $2x$? That would be $2x - 7$.

<u>Answer</u>: a. **No, this is not a math sentence; it's an expression.**
b. $2x - 7$

2. Seven is less than twice x

3. 13 is greater than triple c

4. 12 greater than triple c

5. 5 less than half of y

6. 7 is more than one-fourth of w

7. 8 more than one-third of x is 11

(Answers on pp. 320–1)

Label-Makers and Variables

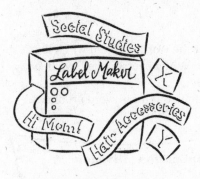

Do you have one of those label-makers? I don't, but I've seen people with them, and they look kinda cool, if maybe a little addictive. You punch in what you want it to say, and it spits out a sticker with your text on it: "Social Studies," "Hair Accessories," "Hi, Mom," etc. These people become very label-happy, making stickers and labeling everything in sight.

And they'd probably make excellent word problem solvers. Here's why: The best way to start word problems is to look for values you don't know and label them!

"Jean is saving for a cute pair of sandals. She is going to add $12 to the money in her bank account." If that is part of a word problem, then you think: "Hmm, what's the *value* I don't know? The amount of *money* that is currently in her bank account." So let's label it! Let's use m for money: Let m = money (in dollars) in her bank account.

And what happens to m? Jean adds $12 to it, so translating: She'll have $m + 12$ dollars. In an actual word problem, you might be asked to *do* something with that, but for now let's not worry about it and just keep on translating.

Here's another example: "Annie's age is doubled and then subtracted from 21." What *value* don't we know here? We don't know Annie's *age*. Let's use a as the placeholder for Annie's age: Let a = Annie's age (in years). Read the problem again and see what *happens* to a? It gets doubled and then subtracted from 21. Let's translate into math-speak:

$$21 - 2a$$

See? That's twice Annie's age, subtracted from 21. Notice that the math sentence might not always be in the same order as the words in the *English* sentence.*

How about this one: "Connor wants to be taller. He said that if he multiplied his present height by 12 and then divided the result by 11, he'd be his ideal height."

In this example, there are two things we don't know: his *present* height and his *ideal* height. So let's use the variables *p* and *i* for our translation. Why not? (If you use *h* for *height* you might be confused about which one you meant.)

So if *p* is his present height, it looks like we want to multiply *p* by 12 (that would be 12*p*) and then divide the entire result by 11. That would be $\frac{12p}{11}$, right? And he says that this is his ideal height (which is *i*, right?), so that suggests an equals sign!

$$i = \frac{12p}{11}$$

Who knows what happens next? Maybe the word problem will tell us what *p* is, or maybe it'll tell us what *i* is. Whichever it does, we'll be prepared, because we've translated correctly.

Here's a summary of some key words to look out for in word problems.

Addition	Subtraction	Multiplication	Division
added to	subtracted from	multiplied by	divided by
sum, total	difference	product	quotient
plus	minus	times	per
more, more than older than additional	less, less than younger than	of *(only when "of" is immediately surrounded by two numbers)* double, twice, triple	a *(only if "a" could be replaced by "per," and the sentence's meaning is unchanged)*
increased by	decreased by	each for every	shared equally split equally

* If you've ever studied French or Spanish, you know that in these languages, adjectives mostly appear *after* nouns, which is the opposite order from English. In French, "brown cat" becomes *chat brun*! The same often happens in math: The *order* of terms might have to change in order to translate the correct meaning.

Of course, each problem is different and you should always use logic when figuring out how to translate into math, but this list can certainly help!

Step By Step

Translating English into math:

Step 1. First, read over the whole problem to get a feeling for what's going on.

Step 2. Look for the value you *don't know*, and label it with a variable of your choice. If there's more than one unknown, you can label more than one.

Step 3. Read the problem again, and figure out *what happens* to the variable, if anything.

Step 4. Now translate the English into math. Use the chart on page 152 if it helps. Done!

And... Action! Step By Step In Action

Some fraction is $\frac{1}{2}$ of $\frac{2}{3}$.

Hmm. This doesn't really look like part of a word problem, but it sure looks like it needs *translating*. We'll do this one step at a time without trying too hard to figure out what it *means*, because I don't know about you, but these kinds of problems can make me feel dizzy!

Step 1. Reading it, it seems like we'll want to find some sort of fraction. Hmm.

Step 2. Okay, the value we don't know is "some fraction," so let's call that *f*.

Steps 3 and 4. Nothing really seems to happen to the fraction so let's write out the sentence we see, translating the English words into math. The word "of" is *immediately* surrounded by two numbers, so:

Answer: $f = \frac{1}{2} \times \frac{2}{3}$

We've successfully translated the English into math. Great! This one is pretty easy to finish off, so let's go ahead and solve it:

$$f = \frac{1 \times 2}{2 \times 3} \rightarrow f = \frac{2}{6}$$

Then we reduce to get $f = \frac{1}{3}$. We could even translate the original problem *back* into English. Try saying this out loud: "$\frac{1}{3}$ IS $\frac{1}{2}$ of $\frac{2}{3}$." (Warning: This does not make for good party conversation.)

Oddly enough, this makes sense. Imagine having $\frac{2}{3}$ of a pie. Now imagine it being cut into two, big $\frac{1}{3}$ pieces. With this in mind, what is half of that pie? Just one of the pieces. So $\frac{1}{3}$ really is *half* of $\frac{2}{3}$.

Doing the Math

Translate into math expressions. I'll do the first one for you.

1. Kim started off with a lot of bracelets. After buying 3 additional bracelets, she decided to share them all equally between herself and 9 friends for the day. Let $s =$ the number of bracelets she started off with. Write an expression for how many each friend got.

Working out the solution:

Step 1. Reading it, we can tell that Kim is going to divide up a bunch of bracelets, and we might also notice that there will be 10 people total getting the bracelets: 9 friends plus Kim.

Step 2. Hmm, well, what *value don't we know?* We don't know how many bracelets she started off with, right? Let's label it! Let s = number of bracelets she *started off with.*

Steps 3 and 4. What *happens* to s? Well, first Kim adds 3 to it: $s + 3$. Then Kim shares the whole thing equally among herself and 9 friends, which is 10 people total, right? That means that it gets divided into 10 equal parts: $\frac{s+3}{10}$. That's an expression for how many bracelets each friend got. Done!

2.   After doubling the money in her bank account, Trudy spent $5 on music downloads. Let s = the money (in dollars) she started out with. Write an expression for how many dollars she has now.

3.   A fraction is $\frac{1}{4}$ of $\frac{4}{5}$.

4.   Brittany loves frozen red grapes. They taste like candy! She has a whole bowl of them. She eats 5 frozen grapes and then shares the rest equally among herself and her 5 friends—Anne, Nicole, Aliza, Paul, and Kirsten. Let s = the number of grapes she started out with. Write an expression for how many Nicole got. *(Hint: This is very similar to #1.)*

5.   Chris had way too many text messages on his phone, and wanted to clear them out. He first deleted 10 text messages, but he still had way too many, so he deleted half of what was left. Let s = the number of text messages he started out with. Write an expression for how many text messages are in his phone now.

6.   Sarah had a lot of ringtones in her phone yesterday. But she kept adding to them, and today she has 9 more than twice what she started out with! Let y = number of ringtones she had yesterday. Write an expression for how many ringtones she has now.

7.   Suzanne works at a pet store. During the morning of her last day working there, she had lots of puppies sitting in the playpen. By the end of the day, she'd sold $\frac{4}{5}$ of them. Of those

remaining, she took 2 home to keep for herself. Let m = the number of puppies she had in the morning. Write an expression for how many are left in the store after Suzanne goes home. *(Hint: Do this one step at a time!)*

(Answers on p. 321)

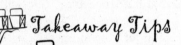

Takeaway Tips

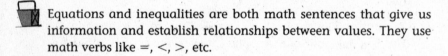

Just like English, math is a language that uses verbs and sentences to communicate information!

Equations and inequalities are both math sentences that give us information and establish relationships between values. They use math verbs like =, <, >, etc.

Expressions are not math sentences; they don't give us any information, and they do *not* use math verbs.

In word problems, your first job is to translate the English into math, and a great way to start is by labeling the unknown values.

The other key to translating word problems is to write down (in math language) any *relationships* that you can between the variables and values.

Danica's Diary

WHY ISN'T HE CALLING?
(OR EMAILING OR TEXTING?)

I know the agonizing pain of waiting for a
guy to call. This happened to me again and
again, from the age of 13 through 17, after
which point I got my first "real" boyfriend,
who actually loved me and *wanted* to call me. We were
together for four years! Sure, once I was single
again, there were still a few guys whose call I
waited longingly for. But somehow, after my first
real relationship, I had more confidence in myself,
and I didn't let it eat me up inside the same way it
did when I was 14.

So, I had the biggest crush on a guy. It was
right before Christmas break, and I remember he said
he was going to "call me." Well, I waited and waited
for the phone to ring, and each time it did, my
heart would leap: THIS COULD BE HIM. More than a
week had gone by, and no phone call from him. I
remember decorating the Christmas tree with my
family, and *all* I wanted for Christmas was for this
guy to call me. I had difficulty enjoying anything
about the holiday; in fact, I only enjoyed it when I
temporarily forgot about whether he was going to
call!

And then . . . the phone rang. And it was HIM!!! I
can't believe I didn't pass out from excitement, but
somehow, I was able to speak in full sentences (at
least that's what I'd like to believe). The
conversation was very short, but I loved it. He
wished me a Merry Christmas, we chatted for a
moment, and then we got off the phone.

After the holidays, nothing ever happened between
us—never a kiss, nothing. To be honest, a part of
me thinks my mom actually called his mom and
arranged for that phone call (they knew each other).
But isn't it sad that my happiness depended
completely on whether some boy called me or not?

I'll tell you why it meant so much to me. When he

said, "I'll call you," I heard this: *I'll call you because I think I'm in love with you. I want to kiss you, and I've been thinking about maybe spending the rest of my life with you. Will you marry me?*

I wish I were exaggerating. But somewhere deep inside, that's what I had decided he was thinking or would be thinking soon. So how could he not call? And if he didn't call, maybe he changed his mind about, you know, the wedding in 10 years! I felt I was going to *lose* so much if that phone didn't ring.

But none of that hinged on the phone ringing. As you now know, the phone did ring, and I got exactly what the ringing promised: a phone call. No more, no less.

I've since learned that it's challenging enough to translate guy-speak as it is, so we shouldn't spend too much energy trying to think of what guys might be *thinking*, too. All those hours on the phone with our friends can be fun, but sometimes they're a breeding ground for reading too much into what a guy has said . . . and we end up jumping to the wrong conclusions—and not enjoying the eggnog. (By the way, do you know what's better than trying to translate guy-speak? Noticing their *actions*—they're much more telling.)

When I'm just dying to know something—whether it's about a guy or an offer on a house or an acting role or anything else I'm "waiting by the phone" for—I've learned to sit back, relax, look at the information I *actually* have (which usually is not very much), and tell myself, "There's nothing I can do about this right now, and worrying or guessing the outcome would be a waste of time, because *there's more information to come.*" And then I go do something else to take my mind off it. With a little practice, you'll find it really works!

More Than 20 Ways to Beat Stress!

Are you dealing with STRESS from homework, grades, or anything else? Sometimes it doesn't take much to help: Music, exercise, or even just a deep breath can make you feel better. In these quotes, you'll find more than 20 ways to beat stress. Pick some and give 'em a try!

"I like to listen to music—it always helps when I'm stressed out."
Keara, 16

"I look at some of my favorite pictures, and it automatically sends this unstressed vibe through me that makes me feel happy inside!"
Robyn, 13

"When I get stressed over school, friends, or life, I grab my iPod and go for a long run." **Stephanie, 14**

"When I have stress, I take a deep breath, tell myself I can do it, and try again." **Amber, 14**

"To get rid of my stress, I take it out on my pillow, go for a walk, or clean. I also like to go work in the yard. That helps, too." **Amanda, 15**

"I cry, or pray, or talk to friends. It depends on the type of stress. I might even do all three. They all make me feel better."
Abby, 14

"I hold it in and talk to my parents about it." **Corey, 14**

"To handle my stress, I usually do the hokey-pokey. It sounds funny and weird, but when I 'shake it all about,' it's like I'm shaking out my stress." **Cheyenne, 14**

"I handle stress by taking a deep breath and then calmly doing the stressful task." **Megan, 13**

"I deal with stress by turning on the radio and dancing or getting out my list of cheers and practicing my cheerleading." *Marisa, 14*

"I have a lot of stress during the school year, especially right after school. If I don't have too much homework, I go to my room and settle down with a book and read." *Tori, 13*

"I play sports or just watch funny movies to forget about stress. But I know I have to get back to my homework soon, because otherwise I'm just procrastinating, and when I get behind on projects, I get even more stressed out!" *Jessica, 16*

"I handle stress in many different ways, but in math, mostly when I have a test, I take a deep breath and then I am usually fine." *Melanie, 14*

"I handle stress pretty well, but sometimes I just need to scream into a pillow." *Grace, 14*

"I get stressed out really easily. Sometimes I journal to get my feelings out." *Jenna, 18*

"When I have a lot of homework, to make sure I don't get stressed, I usually do one subject at a time, and in between, I take small breaks, maybe get a snack and watch a little TV. I am usually pretty good about handling stress." *Hannah, 14*

"I'm a dancer, and if I'm stressed out at home or at school, when I go to dance, I can just dance my stress away." *Rachel, 14*

"I like to cook—it just melts my stress away!" *Dawn, 17*

Math . . .
In Jobs You Might Never Expect!

Knowing math can benefit you in a variety of jobs you might never imagine. I mean, who'd figure a lawyer would need to know math? (See p. 277.) And here are a bunch more jobs that use math! Do any of these jobs sound like *you*?

Actress: In the highly competitive world of entertainment—whether it be the bright lights of Hollywood or the glamour of the New York theater—being an actress is more than just performing. The camera fails to capture the "business" in show business! We typically will give 10% of our salary to the agent, 10% to the manager, and 5% to the lawyer, plus the publicist gets a flat fee, which needs to be budgeted for. Savvy actresses benefit from being able to read and understand the math in the contracts, or these people might rip them off! (And you can bet it happens, too.) I speak from experience when I say that a sharp brain is needed to memorize lots of lines, especially the rewrites that come in the night before, and math helps keep my brain sharp.

Interior Designer: Are you addicted to those home decorating shows where they turn an old shack into a chic little apartment? Do you find yourself pushing furniture around in your bedroom, rearranging chairs, bookshelves, and paintings? As an interior designer, an empty room is your canvas! Though it plays a large part, creativity is not the *only* requirement for being a great interior designer. According to New York interior designer Cat Lindsay: "Math is used more often than one might think . . . furniture layouts require dimensioning (this involves adding fractions). To obtain square footage of floor plates, we have to calculate areas of triangles, parallelograms, and other geometries. When designing for law firms, calculations are also required when comparing ratios—for instance, how many secretaries there are per lawyer—so we can figure out how to best arrange the offices. Also, if you want to design a fabulous curved sofa or chair, lots of geometry is needed!"

Veterinarian: If you're like most people, you consider your pet to be a part of the family. Besides making excellent companions, animals often have therapeutic value for their owners. If you love animals, a career as a veterinarian might be for you! Choosing to be a veterinarian is a serious decision, but the rewards are countless. And there's no question that you'll need math: first to get your degree and then to care for the animals. According to Los Angeles veterinarian Tina Chang: "I love spending my days helping animals. Our job includes treating disease, alleviating pain and suffering, and saving lives. With that said, if we do not perfect our math skills, we can make mistakes and cause harm! Being a veterinarian requires the use of mathematics on a daily basis, from simple tasks like weighing an animal to more complex ones like using metric system unit conversions and math formulas to determine appropriate doses of lifesaving medication."

Event Planner: Have you ever been to a wedding or a huge party where everything seems perfect—right down to the candles, the music, and the flowers? If you're the kind of girl who loves to arrange parties as much, if not more, than actually attending them, then event planning might be a good match for you, and it takes more math than you might guess. According to New York City event planner Allison Lafferty: "We use math all the time to calculate things like the square footage of the space where the event takes place to ensure all of the tables, food, and DJ booth will fit. We also use math to determine the amount of decorations and food/beverage needed for the number of attendees. (For example, one gallon of coffee usually fills 15 cups, so depending on the time of day, we first estimate how many people will drink coffee.)" *Chicago Social* magazine event planner Kelly Berg adds: "As an event planner, I'm given a budget that I must work within, which means I need to track expenses for individual events and entire fiscal periods. For example, how much did I spend on catering and event décor at my past 10 events? What percent of the event décor cost was dedicated to flowers? Yep, behind every good party there are proportions, fractions, integers, you name it!"

Architect: It's easy to go about our daily lives, moving in and out of houses and buildings without imagining how they were actually built. But, when you think about it, it's pretty amazing! The Empire State Building, the Jefferson Memorial, and the Golden Gate Bridge in San Francisco are all astounding, but consider the Seven Wonders of the World: Each is the product of an architect—many architects, and, in some cases, generations of architects. If only all of us could visit the pyramids, the Taj Mahal, or the Great Wall of China, we might have a better appreciation for architecture. It's truly a remarkable art form, but remember that architecture is a *science*, as well. According to New York City architect Michelle Drollette: "To be a successful architect, one needs to be both creative and methodical at the same time. Developers (the people who hire us) have complex pricing models that use square footage (as well as other criteria) to estimate how much profit will be made once they sell the building they're asking us to design. So, we often face the task of designing a space within a narrow square footage requirement while thinking outside the box to create visually interesting spaces. Fulfilling this requirement while we're designing can be like doing an algebra problem with square footage, and a puzzle that depends on creative input, all at the same time!"

Fashion Designer: Do you pore over the latest issue of *Vogue* as soon as it reaches your doorstep? And, besides fantasizing about wearing the clothes you see inside, do you imagine yourself actually *designing* them? If so, maybe being a fashion designer is your true calling. According to couture designer Jessica Wade of JesWade designs: "Fashion designers need to know math for lots of reasons. If [a designer] creates her own designs, then she needs to understand geometry and math for manipulating fabric shapes and matching each seam: Flat planes of material must become three-dimensional masterpieces! And each final design has to be fitted, which means additional measurements and proportions when making clothing for all the different sizes. Finally, fashion is a business, like anything else: A designer needs to budget the cost of clothing materials and be able to analyze sales reports to see how her clothing is selling!"

Gourmet Chef: Behind any great dining experience, there is a great chef! With waiters rushing in and out, steaming pots on every burner, and cooks scrambling to get each order in on time, the kitchen makes up the nerve center of the restaurant. And, as chef, you run the whole show! Maybe you create your own delicious recipes too—rare entrees, special sauces, or even those elegant desserts they roll out on carts for the customers to drool over! According to chic, four-star, New York City restaurant *Le Bernardin* chef Stacia Woodrich: "To be successful in a professional kitchen, a chef must understand fractions and proportions in order to adjust recipes for the specific number of people we are catering to." Yep, fractions and proportions are helpful in the kitchen, whether you're adjusting a cookie recipe at home for friends or serving large groups of people in a swanky French restaurant.

Film Producer: Are you a huge film buff? Ever wonder what some of your favorite movies look like from behind the scenes? Well, you can be sure there's a lot more going on than what you see on the screen! Film producers, for example, are part of the movie even before the actors and actresses are, and yes, they are in charge of the financial aspects of a film. According to Hollywood film producer Kim Zubick: "Film producers use tons of math: from estimating costs and meeting studios' budget requirements to get the green light, to negotiating salary rates (which can also involve percentages of profits) and the schedules and budgets themselves (often involving ratios and algebra). Film budgets are truly math puzzles that keep evolving while we shoot the movie. It's the producer's job to keep the movie on budget, or the movie might never make it to the big screen!" (Also, check out TV director/producer Pam Fryman's Testimonial on p. 190.)

Doctor: To be a doctor is not only an enormous accomplishment, it is probably one of the most admirable professions there is. As a doctor, maybe you'll perform lifesaving surgery on a sick patient or travel to third-world countries to give much-needed medical care. You might experience the thrill of delivering a baby or helping an accident victim walk again. According to fabulous New York OBGYN Dr. Laura Meyer:

"Several mathematics courses were required in college. And now I use math almost every day, whether I'm calculating a medication dose based on a patient's body weight or plugging lab values into an equation to evaluate a patient's kidney function." She also says, "One of the most exciting parts of my job is delivering babies!"

Boutique Owner: What's great about an independent boutique is that most of the clothing and accessories shoppers find there are unique. They don't necessarily follow all the current trends. Are you the type of girl who chooses your wardrobe based on your own individual style? Maybe you'd be interested in opening your own small store where all the items are personally selected by you. According to Eve Newhart, owner of chic Los Angeles boutique Wicati: "Math is just as important to my success as fashion sense! Owning a boutique is all about the numbers—how much to order, how much to spend, how much to mark down, how much to project for the future, and more. I love owning a boutique clothing store, and believe me, math keeps the doors open."

Photographer: How many times have you said to yourself, "If only I had my camera with me right now"? Well, it is the job of the photographer to *always* be ready to capture that amazing sunset over the water, or to document the events of a war, a protest, a rally, or even the daily lives of people most of us will never meet and places we will never see. Each photographer has a unique perspective on the world and, in pictures, allows us to see and feel what he or she does. According to Los Angeles–based professional photographer Cathryn Farnsworth: "Photography is mostly creative, but we do use math, too. For example, the intensity of light on a subject decreases proportionally to the square of the distance between the light and the subject. In other words, if a subject is moved 3 times closer to the light, the subject will look 9 times as bright!"

***

To hear from two more lovely ladies who also use math in their jobs—my sister and my best friend—see pp. 277–8!

Chapter 12

The Art of Gift Wrapping

Solving Equations

𝒩 ow that we've honed our talents with variables, it's time to use those skills to get really good at solving for *x*, which has a little something to do with . . . the holidays. (Just go with me here.)

I wonder if the holidays leave dogs rather perplexed. They watch us do things like hide painted eggs in the yard in springtime, watch fireworks in July, stick candles in food and then light them and blow them out, dress up in costumes in October, and bring a tree inside to decorate.*

As if this weren't confusing enough, Sparky watches us bring presents home and wrap them up, only to *unwrap* them again, save some of the silky ribbons and sparkly bows, and do it all over again at the next event. What's with all the wrapping and unwrapping? I'm guessing he'd just as soon get the toy and be done with it.

If you're like me, shiny wrapping paper and bows are one of the best parts of holidays. Another benefit of wrapping and unwrapping presents? It makes understanding *inverse operations* so much easier!

. . . .

* They totally "get" Thanksgiving, though.

What's It Called?

Inverse Operations

Inverse operations are operations that <u>undo</u> each other. For example: Opening a box and closing a box are inverse operations. They undo each other. Addition and subtraction are inverse operations. Addition undoes subtraction, and subtraction undoes addition.

You may have seen something like this before. It's a list of operations and the operations that *undo* them; in other words, their inverse operations!

Operation		How to Undo It (its *inverse* operation)
Addition	$\leftrightarrow$	Subtraction
Subtraction	$\leftrightarrow$	Addition
Multiplication	$\leftrightarrow$	Division
Division	$\leftrightarrow$	Multiplication
Squaring	$\leftrightarrow$	Taking the Square Root
Taking the Square Root	$\leftrightarrow$	Squaring

So if we start with plain 'ol x and we add 2, we get $x + 2$. *Undoing* that action would mean subtracting 2, right? That looks like this: $x + 2 - 2 = x$. We end up where we started, because we *undid* the adding of 2.

How about if we start with x and multiply it by 3? So we'd do $x \rightarrow 3x$. To *undo* that action and get back to where we started, we'd divide by 3, right? So, $3x \rightarrow \frac{3x}{3} = x$. Obviously, this works for numbers, too.

If we start off with 5 and square it, we get $5^2 = 25$. Then, to undo what we've just done, we would take its square root, $\sqrt{25} = 5$, and we're right back to where we started. You probably won't use square roots much to solve equations until later in algebra, but I wanted to show you how taking square roots and squaring are inverse operations, too!*

Gift Wrapping 101

We've talked about undoing *single* operations so far; it's just a one-step "undoing" process. The process of undoing *more* than one step (like in solving equations) is just like unwrapping a gift . . . or taking off boots!

Let's say you put on some ankle socks and then some cute boots. To undo this, you'd have to *first* take off the boots and *then* take off the ankle socks, obviously. So if you did A and then did B, to UNDO what you've done, first you'd undo B and *then* you'd undo A. Makes sense, right? Now let's see how it works when there are *three* steps.

If you wanted to wrap a beautiful pink sweater for your sister, first you'd put it in a box. Then you'd wrap the box in wrapping paper, and then you'd stick a sparkly bow on it. When your sister unwraps it, she would do the inverse of each action you did, in the reverse order.

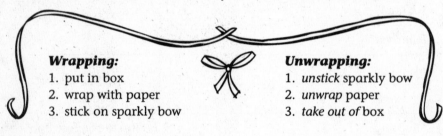

Wrapping:
1. put in box
2. wrap with paper
3. stick on sparkly bow

Unwrapping:
1. *unstick* sparkly bow
2. *unwrap* paper
3. *take out of* box

See how the *first* thing she does to unwrap the gift undoes the *last* thing you did when you wrapped it? And how the *last* thing she does to unwrap it undoes the *first* thing you did to wrap it? (Let's just assume she takes the time to remove the bow, okay?)

And believe it or not, when we isolate *x* by undoing a series of operations, it works just the same way!

. . . .

* Technically, "squaring" and "taking the square root" are only *inverse operations* for positive numbers (or zero). You can't wrap and unwrap a negative number with those operations and end up with the same thing again. For example, if you start out with -5 and square it, you get $(-5)^2 = 25$. But $\sqrt{25} = 5$, not -5. This footnote is pretty advanced, so don't worry if you don't understand it. You'll learn this later in algebra.

What the Movie Stars Are Saying!

"*I* feel so gratified to have finished college. I learned how to articulate myself. It gave me confidence more than anything."
Maggie Gyllenhall (actress, Mona Lisa Smile, Stranger Than Fiction)

Isolating *x*

Let's take a look at this: $2(x + 3) - 8$. How did this come to be?

Once upon a time, *x* was all by himself. Then someone wrapped him up! Here's what happened to him: First, 3 was added to him: $x + 3$. Then the whole thing was multiplied by 2, and here's how he looked at that point: $2(x + 3)$. THEN, 8 was subtracted from this whole thing, so this is how he looks now: $\mathbf{2(x + 3) - 8}$.

He can hardly recognize himself. Let's get him out of there! In order to unwrap *x*, first we'd need to add 8, and we'd get $2(x + 3)$. Then we could divide the whole thing by 2 and get $x + 3$. And then, subtracting 3, we'd finally get *x* back to his normal self again, totally unwrapped. Nice.

See how isolating *x* has more to do with the holidays than you might have thought?

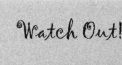

Watch Out!

Notice how I keep saying "the *whole thing* was multiplied by 2" or "then we subtract 8 from the *whole thing*." When we're wrapping *x* or unwrapping *x*, it's important that we always "do" things to the entire expression, not just one part of the expression. So if we had $2x + 1$ and we wanted to wrap it up more by dividing by 2, we could NOT just divide the $2x$ by 2; we'd have to divide the *whole thing* by 2: $\frac{2x + 1}{2}$.

This is because soon we'll be using these techniques to solve for *x*, and when you do things to both sides of the equation, you need to do things to each *entire* side of the equation in order to keep the scales balanced.

QUICK NOTE   I keep saying x, but of course this works for any variable you want to isolate: a, b, c, n, w, x, y, z, □, ☺, ✿, etc.

Before we get into solving for *x*, let's practice the skill of wrapping and unwrapping *x*!

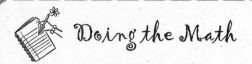

Doing the Math

I'll describe the steps that were used to build an expression, starting with *x* (or some other variable). Your job is to actually build it! Remember at each step to do "things" to the *entire* expression. Then, list the *unwrapping* steps. I'll do the first one for you.

1.   Start with *y*. Divide by 8, then subtract 4, and then multiply by 3.

<u>Working out the solution</u>: Okay, we start with y, and we divide by 8. That can be written like this, $\frac{y}{8}$, right? Next, we're supposed to subtract 4. Remember, this would be wrong: $\frac{y-4}{8}$. We have to subtract 4 from the *whole thing*, so we would write $\frac{y}{8} - 4$. So far, so good?* Now we're ready for the next instruction: "Multiply by 3." We know we must multiply the *whole thing* by 3, so that means $3\left(\frac{y}{8} - 4\right)$. Okay, we're done with that part! The unwrapping steps would be the *inverse* of the first instructions we got, just like unwrapping a gift: Divide by 3, add 4, and multiply by 8. Done!

• • • • • • • • • •

* BTW, you would build $\frac{y-4}{8}$ by first subtracting 4 from *y* and *then* dividing by 8.

<u>Answer:</u> $3\left(\dfrac{y}{8} - 4\right)$. And to unwrap it, we'd first divide by 3, then add 4, and then multiply by 8.

2. Start with x. Add 3, and then multiply by 4.

3. Start with y. Multiply by 4, and then add 3.

4. Start with z. Add 3, and then divide by 4.

5. Start with w. Divide by 2, then subtract 1, and then multiply by 5.

6. Start with n. Multiply by 6, then subtract 5, and then divide by 7.

(Answers on p. 321)

Solving for x

Now let's apply our unwrapping knowledge to solving for x. Before we do, let's review what it *means* to solve equations. Remember, when you get an equation to solve like $5(2x + 1) - 6 = 29$, you're being asked to find out what number x has to be in order for the equation to indeed be a true statement. That's our job—to discover x's value! Sure, we could just stick a bunch of values in until one of them works, but x is often a fraction. Are you really going to guess every fraction you can think of, too? There's got to be a better way to find out x's value, and there is!

Mini Review* of **Intro to Solving for x**
as Taught in Math Doesn't Suck

If you think of *x* as a bag of pearls, then when we solve for *x*, our job is to find out *how many pearls* are in that bag. The way we do this is to think of the equation as a set of balanced scales. Let's look at a simple equation:

$$2x + 3 = 13$$

This is a true statement; the scales are balanced. Now, if we do the *same things* to both sides of the scale, we will keep the scales balanced. In this case, we'd take away 3 loose pearls from both sides, leaving 2 bags on the left and 10 pearls on the right: $2x = 10$. At this point, we could take away half from both sides (divide both sides by 2) and end up with one bag on the left and 5 pearls on the right: $x = 5$. And voilà! We've figured out how many pearls are in the bag. In other words, we've figured out the value of *x* that makes the statement true by keeping the scales balanced the whole time.

> **When a math problem asks you to find the value of *x*, the goal is to isolate *x* by doing things to both sides of the equation until *x* is all by itself on one side and a number is on the other side. That number is your answer!**

. . . .

* For a more in-depth review of the basics of solving for *x*, check out Chapter 20 in *Math Doesn't Suck.*

When you have a more complicated equation like $5(2x + 1) - 6 = 29$, it's a little harder to draw out the bags and pearls. So let's find a better way! *(Hint: It has to do with gift wrapping.)*

In the DOING THE MATH on pp. 170–1, we practiced wrapping (building expressions) and then unwrapping *x* in order to isolate *x* again. That skill is about to come in handy. Now, instead of just building an expression, we're going to build a whole equation!*

Building an equation ⟷ Wrapping a present

Solving an equation ⟷ Unwrapping a present

My favorite number is 3, so let's decide that the final answer will be **x = 3**, and build an equation from that. Let's wrap!

Building an Equation: Wrapping Up x

$x = 3$	Once upon a time, x was alone on one side of the equation.
$2x = 6$	So we multiplied both sides by 2.
$2x + 1 = 7$	Then we added 1 to both sides.
$5(2x + 1) = 35$	After that, we multiplied both sides of the equation by 5.
$5(2x + 1) - 6 = 29$	And then we subtracted 6 from both sides. I think we've done enough wrapping, don't you?

• • • • • • • • • •

* To review the difference between *expression* and *equation*, see pp. 146–7. Briefly: An equation uses the = sign, and an expression does not.

And we've created the following equation, which we secretly know will be true for $x = 3$:

$$5(2x + 1) - 6 = 29$$

Ta-da! All wrapped up and ready for someone else to solve.

QUICK NOTE   Notice that during the wrapping, at the stage where we had $2x + 1 = 7$ and decided to multiply both sides by 5, we *had* to use parentheses to surround the $(2x + 1)$. This is because when you do something to both sides of an equation, you must do that same thing to each *entire* side of the equation, not just the part touching the variable. That's how you make sure to keep the scales balanced, which is absolutely the most important part of solving for x.

Now let's see how unwrapping x from $5(2x + 1) - 6 = 29$ compares to the wrapping process we just did. Remember, just like when we unwrap a gift, in order to isolate x, we need to do the inverse of the <u>last</u> step . . . *first*. And then we keep going, dealing with each new layer as we get to it—all with the goal of getting x by itself on one side. Let's do it!

Solving an Equation: Unwrapping x

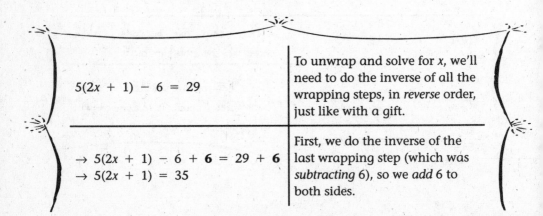

$5(2x + 1) - 6 = 29$	To unwrap and solve for x, we'll need to do the inverse of all the wrapping steps, in *reverse* order, just like with a gift.
$\rightarrow 5(2x + 1) - 6 + \mathbf{6} = 29 + \mathbf{6}$ $\rightarrow 5(2x + 1) = 35$	First, we do the inverse of the last wrapping step (which was *subtracting* 6), so we *add* 6 to both sides.

$\rightarrow \dfrac{5(2x + 1)}{5} = \dfrac{35}{5}$ $\rightarrow \dfrac{\cancel{5}(2x + 1)}{\cancel{5}} = \dfrac{^7\cancel{35}}{\cancel{5}}$	The second-to-last step in wrapping was to *multiply* by 5, so now we need to *divide* both sides by 5. The factors of 5 cancel nicely, don't they?
$\rightarrow 2x + 1 = 7$ $\rightarrow 2x + 1 - \mathbf{1} = 7 - \mathbf{1}$	Next, we do the next inverse and *subtract* 1 from both sides. (When we wrapped, we *added* 1). We're almost there!
$\rightarrow 2x = 6$ $\rightarrow \dfrac{2x}{2} = \dfrac{6}{2}$	For the last step in unwrapping, we need to *divide* both sides by 2, and when the 2's cancel, we'll have x all by itself. (This makes sense because the first thing we did in wrapping was to *multiply* by 2.)
$\rightarrow x = 3$	And voilà! We end up with x on one side, totally unwrapped and revealing to us its identity, once and for all.

Notice that at each step along the way while we're solving, we keep the equals sign—this is because we have true statements of equality at *each stage.*

Most of the time, you won't be unwrapping a present you wrapped yourself (or solving an equation you built yourself). But now, you've seen what's going on behind the scenes with equations, and you're ready for an unwrapping *shortcut*! If you get confused about the correct order in which to unwrap things, you can think of it as *undoing PEMDAS.*

Shortcut Alert:
Order of Unwrapping *x* ⟷ Undoing PEMDAS

Remember the order of operations (and the pandas) from p. 21? PEMDAS: Parentheses, Exponents, Multiplication and Division together, Addition and Subtraction together. This is the order in which we must always simplify expressions. As it turns out, when we unwrap *x*, we are *undoing* PEMDAS. Let's take a look at just the *left-hand side* of the equation we wrapped and unwrapped on pp. 173–5:

$$5(2x + 1) - 6$$

Notice that when we unwrapped to find *x*, we followed the *reverse* order of PEMDAS. By first undoing the subtraction, and THEN undoing the multiplication of 5, and finally getting to the inside of the parentheses, we were *undoing* PEMDAS. So, the unwrapping order is: Addition and Subtraction together, Multiplication and Division together, then Exponents, and then Parentheses.

QUICK NOTE   Since the *last* thing we do in PEMDAS is addition and subtraction, if you notice any lone addition or subtraction hanging out, it's going to be the *first* thing that needs to get undone! (This, of course, doesn't apply to addition or subtraction that happens inside Parentheses.)

Notice that in a situation like this, $\frac{3(x + 1)}{2} = 9$, you have choices. There is no addition or subtraction outside of the parentheses, so the next thing on the undoing PEMDAS list would be to undo multiplication and division, right? And because multiplication and division have the same priority, you can either multiply both sides by 2 first, or you can divide both sides by 3 first. It's your choice.

Solving for x (in equations with one x):

Step 1. Try to figure this out: What was the *last thing* done to wrap x like this? Undo that action first. Remember that you're *undoing* PEMDAS.

Step 2. Continue to unwrap, one step at a time, with the goal of getting x by itself on one side of the equation and numbers on the other side. Make sure to do the same thing to both sides of the equation until x is by itself.

Step 3. Once you find x's value, substitute it back into the original equation and make sure you get a true statement. This is an easy way to make sure no careless mistakes were made and to ensure you get the homework/test points you deserve!

 *And... Action!* Step By Step In Action

---

Let's solve $-2(x - 4) + 3 = 9$.

Step 1. Using the undoing PEMDAS shortcut, let's first subtract the 3 from both sides. (Remember: The "-2" is being *multiplied* here, not subtracted.)*

$$-2(x - 4) + 3 - 3 = 9 - 3$$
$$\rightarrow -2(x - 4) = 6$$

Step 2. Continuing, how do we peel off the next layer of wrapping? Looks like if we divide by -2, we'll be able to undo the multiplication of -2.

$$\rightarrow \frac{-2(x - 4)}{-2} = \frac{6}{-2}$$

The -2's will cancel on the left side of the equation. For the right side, we know from integer division (see p. 43) that 6 divided by -2 will give us -3, right? So:

$$\rightarrow (x - 4) = -3$$

. . . .

* Even if the order of the problem were switched around, and you couldn't see a plus sign, $3 - 2(x - 4) = 9$, you could still subtract 3 from both sides first. In fact, this equation and the one above are actually the same equation. (This is harder to conceive, I know, which is why it's only in a footnote!)

Wow, that's looking *much* better. There's no reason to keep the parentheses anymore, now that the expression on the left isn't getting multiplied by anything, so we can rewrite this as $x - 4 = -3$.

Let's now add 4 to both sides (which must have been the first thing that wrapped up poor little *x*), and we'll get:

$$x - 4 + 4 = -3 + 4 \rightarrow x = 1$$

Step 3. To check our answer, let's substitute the value $x = 1$ into the original equation:

$$\overset{?}{-2(1 - 4) + 3 = 9} \rightarrow \overset{?}{-2(-3) + 3 = 9} \rightarrow \overset{?}{6 + 3 = 9} \rightarrow 9 = 9 \checkmark$$

Yep, we got a true statement, so we found *x*'s true value.

Answer: $x = 1$

Watch Out!

Going quickly, we might have accidentally written $x = 9$ as our answer because we got $9 = 9$ as our true statement when we *checked* the answer $x = 1$. Just write things out clearly, keep your brain on, and you'll be fine!

Notice that we could have started this problem by distributing* the -2 inside the parentheses to the *x* and the -4. So, we could have rewritten it like this (notice how the negative signs cancel to give us positive 8):

$$-2(x - 4) + 3 = 9$$
$$\rightarrow -2x + 8 + 3 = 9$$
$$\rightarrow -2x + 11 = 9$$

And we could have continued from there. Distributing first is sometimes an option (which might make your work easier). You'll get a totally equivalent equation!

• • • • • • • • • •

* To review *distributing* and how it interacts with PEMDAS, see Chapter 10.

Doing the Math

Unwrap these equations and solve for x. Find out what value of x will make these statements true. When you do, plug this value back in to check your answer. I'll do the first one for you.

1. $\frac{(x + 2)}{5} + 1 = 3$

Working out the solution: It looks like the 1 was added last, so let's subtract 1 from both sides: $\frac{(x + 2)}{5} + 1 - 1 =$ $3 - 1 \rightarrow \frac{(x + 2)}{5} = 2$. Continuing to unwrap, it seems we can get rid of the division by 5, by *multiplying* both sides by 5: $\frac{5(x + 2)}{5} = 5(2)$. The 5's on the left will cancel just like we planned it, so we get: $(x + 2) = 10$. We can now safely drop the parentheses and then subtract 2 from both sides: $x + 2 - 2 = 10 - 2 \rightarrow x = 8$. We've been doing the same thing to both sides of the equation, and now we are rewarded with finding out x's value! Plugging in $x = 8$ to see if we get a true statement: $\frac{(8 + 2)}{5} + 1 = 3$

$\rightarrow \frac{10}{5} + 1 = 3$

$\rightarrow 2 + 1 = 3$

$\rightarrow 3 = 3$ Yep!

Answer: $x = 8$

2. $2(x - 6) = -18$

3. $\frac{(x - 4)}{2} = 1$

4. $3(x - 5) - 2 = 7$

5. $\frac{(x + 1)}{3} + 2 = 3$

(Answers on p. 321)

What Do You Really Think?

In an anonymous poll, we asked more than 200 girls ages 13 to 18 the following question, and here's what they said!

When you first struggle with a math problem, your first instinct is to . . .

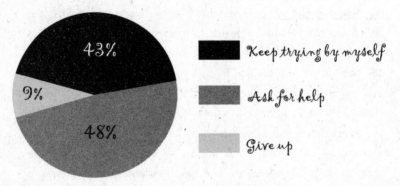

- 43% *Keep trying by myself*
- 48% *Ask for help*
- 9% *Give up*

My advice is, when you first struggle with a problem, *keep trying*. Then, if you really don't get it, ask for help. Whatever you do, don't give up!

Solving for *x* When There's More Than One *x*

Sometimes you'll need to solve an equation that has more than one *x* in it, like this:

$$8x + 4 = 15 - 3x$$

It would be nice to get rid of that $-3x$ on the right side, wouldn't it? Sure enough, the first thing we do is collect all the *x* terms together on one side. How can we collect them? By doing the same thing to both sides of the equation, like always! If we add $3x$ to both sides, we get:

$$8x + \mathbf{3x} + 4 = 15 - 3x + \mathbf{3x}$$

Combining like terms, just like we did in Chapter 9, we get:

$$11x + 4 = 15$$

And then we can continue as normal! (The answer is $x = 1$, by the way.)

Step By Step

Solving for x (when there's more than one x):

Step 1. Collect all the "stuff with variables" on one side by adding and subtracting things on *both sides* of the equation and sometimes using the distributive property. Using the same methods, collect all the individual numbers (constants) on the other side.

Step 2. Combine like terms so there is only one *term* with x in it, like $5x$ or $\frac{3x}{2}$.

Step 3. Continue by *undoing* PEMDAS one step at a time until you have just one x by itself on one side (unwrapped!) and just a number on the other side. Make sure you keep the scales balanced at each step.

Step 4. Once you get x's value, always check your answer by substituting it back into the original equation and making sure you get a true statement. Done!

 And... Action!

Step By Step In Action

Solve for x: What value of x will make this statement true?

$$6x + 4 = 2(x + 1)$$

Steps 1 and 2. We must start by getting all the x terms on one side and all the constants on the other. Hmm. In order to do this, it looks like we need to *distribute* that 2 so that we can eventually collect the x's together. Doing that, we get:

$$6x + 4 = 2(x + 1)$$
$$\rightarrow \quad 6x + 4 = 2x + 2$$

To get all the x's on one side, let's subtract **2x** from *both* sides and combine like terms:

$$6x - 2x + 4 = 2x - 2x + 2$$
$$\rightarrow 4x + 4 = 2$$

Step 3. Now that x appears on only one side of the equation, we can continue by *undoing* PEMDAS in order to isolate x. Looks like we need to *subtract* 4 from both sides:

$$4x + 4 - 4 = 2 - 4$$

$$\rightarrow 4x = -2$$

And finally, if we *divide* both sides by **4**, we'll get x all by itself, won't we?

$$\frac{4x}{4} = \frac{-2}{4}$$

$$\rightarrow x = \frac{-1}{2}$$

Step 4. Let's check our answer by plugging the value $x = \frac{-1}{2}$ into the original equation:

$$6x + 4 = 2(x + 1)$$

$$\rightarrow 6\left(\frac{-1}{2}\right) + 4 \overset{?}{=} 2\left(\frac{-1}{2} + 1\right)$$

Kinda messy, but it's not so bad: The left side of the equation becomes $-3 + 4 = 1$. On the right side, we can first combine *inside* the parentheses: $\frac{-1}{2} + 1 = \frac{1}{2}$, right? So, the right side $= 2\left(\frac{1}{2}\right) = 1$. We get the true statement, $1 = 1$, so we must have found the correct value for x. Yesss!

Answer: $x = \frac{-1}{2}$

QUICK NOTE   Remember that $3 = x$ is the same as $x = 3$, so if you're working with an equation and isolating x, sometimes x will stay on the right side instead of the left. And that's fine!

Note to my fabulous readers: Since solving for x problems can raise so many different kinds of issues, I'm going to work out two more examples in tons of detail in the following little Workshop. They both follow the step-by-step method but use slightly different strategies to prepare you for whatever your teacher has in store.

Solving for x Workshop

Workshop Problem #1: Solve for x. What value of x will make this statement true?

$$\frac{x}{2} + 3 = 8 - 2x$$

Step 1. To collect all the *x* terms together, we can add 2*x* to both sides, and we'll get:

$$\frac{x}{2} + 2x + 3 = 8 - 2x + 2x$$
$$\rightarrow \frac{x}{2} + 2x + 3 = 8$$

Step 2. And how can we combine $\frac{x}{2} + 2x$? Well, we *could* rewrite 2*x* as a fraction, $\frac{2x}{1}$, find a common denominator (2), and then proceed with fraction addition, but that sounds time-consuming. Instead, let's multiply *both entire sides* of the equation by 2 and see what happens!

$$2\left(\frac{x}{2} + 2x + 3\right) = 2(8)$$
$$\rightarrow \frac{2x}{2} + 4x + 6 = 16$$
$$\rightarrow x + 4x + 6 = 16$$

No more fraction. Yay!

Step 3. Now we'll continue by combining like terms, $x + 4x = 5x$, and subtracting 6 from both sides:

$$\rightarrow 5x + 6 - 6 = 16 - 6$$
$$\rightarrow 5x = 10$$
$$\rightarrow \frac{5x}{5} = \frac{10}{5}$$
$$\rightarrow x = 2$$

So we got **x = 2**.

Step 4. Plugging $x = 2$ into the original equation gives us:

$\frac{x}{2} + 3 = 8 - 2x$

$$\to \frac{2}{2} + 3 \overset{?}{=} 8 - 2(2)$$

$$\to 1 + 3 \overset{?}{=} 8 - 4$$

$$\to 4 = 4 \checkmark$$

Yes, it checks out! We found the value of x that makes the equation a true statement.

Answer: $x = 2$.

<u>Workshop problem #2:</u> Solve for x.

$$4 - x = -9x - 3 + x$$

Steps 1 and 2. First, we're going to remain totally calm about all those negative signs. And so that we can get in control of them, we're going to rewrite all the subtraction as "adding negatives." And hey, let's also write in the sneaky 1 coefficients to make things easier:

$$\to 4 + (-1x) = -9x + (-3) + 1x$$

Next, let's see how we can get all the x's together. On the right, we can combine $-9x$ with $1x$, and we get $-8x$, right?

$$\to 4 + (-1x) = -8x + (-3)$$

We still need to gather all the x terms together. Hmm, what would happen if we *added* $1x$ to both sides? We'd have to combine $-8x + 1x$ on the right side. On the left side, the $1x$'s disappear!

$$\to 4 + (-1x) + 1x = -8x + 1x + (-3)$$

$$\to 4 = -7x + (-3)$$

That's looking a bit better.

Step 3. Now that all the "x stuff" is on one side, to *further* isolate x, let's *add* 3 to both sides, and oh my, it's suddenly looking much friendlier!

$$\rightarrow 4 + 3 = {}^{-}7x + (-3) + 3$$
$$\rightarrow 7 = {}^{-}7x$$

The last thing we need to do to get x all by itself, is to *divide* both sides by -7, right? Let's do it:

$$\rightarrow \frac{7}{-7} = \frac{-7x}{-7}$$

The -7's cancel each other on the right side. On the left side, remember that when you divide 7 by -7, you'll end up with a negative answer: -1.* So we get $-1 = x$.

Step 4. Let's plug $x = -1$ into the original equation, and get:

$$4 - x = -9x - 3 + x$$
$$\rightarrow 4 - (-1) \overset{?}{=} -9(-1) - 3 + (-1)$$

$$\rightarrow 4 + 1 \overset{?}{=} 9 - 3 + (-1)$$
$$\rightarrow 5 \overset{?}{=} 6 + (-1)$$
$$\rightarrow 5 = 5$$

Yep, we did it right and found the value of x that makes this equation true!

Answer: $x = -1$

. . . .

* Remember that you can also rewrite it like this: $\frac{7}{-7} \rightarrow -\frac{7}{7} \rightarrow (-1)\frac{7}{7}$. See p. 46 for a review of this. Then you can just cancel the 7's, and it's easier to see why you end up with -1.

Solving for x:
The Troubleshooting Guide!

Issue #1: Suddenly *one side* of the equation became blank—nothing was there.

Sometimes you'll be solving and you'll get "nothing" on one side of the equation. That just means it's *zero*! For example, if you had $4x + 2 = x$, and you subtracted x from both sides, you *should* get $3x + 2 = 0$. When you subtract something like $x - x$, just remember to write 0 in its place, and you won't end up with "nothing" ever again! To finish this problem, you'd subtract 2 from both sides, and then divide both sides by 3 and get $x = -\frac{2}{3}$.

Issue #2: Suddenly x disappeared from the equation completely!

Sometimes all the x's subtract away or disappear in some other way. If you've done everything correctly, here's what this means:

• If you end up with a true statement like $1 = 1$, the equation is true for *all values* of x.

  Here's an example: $x + 1 = x + 3 - 2$. Plug in a few values for x. See how it's true for *every* value of x? And if we subtract x from both sides, we end up with $1 = 1$.

• If you end up with an *untrue* statement like $1 = 2$, it means that the equation will never be true no matter what x equals. So there is <u>no solution</u>.

  Here's an example: $x = x + 1$. It doesn't matter what you plug in for x; it will *never* be a true statement. There's no way to check your answer for these, so always reread your work.

Issue #3: There Are Too Many Negative Signs.

One thing you can do when there are too many negatives, especially when x has a negative sign, is to <u>multiply both sides of the equation by -1</u>. On p. 183, we did something very similar when we multiplied both sides by 2 in order to get rid of the fraction. It just makes everything nicer, and it's legal because we're keeping the scales balanced. Just be sure, as always, to multiply the -1 times each *entire* side, and use the distributive property correctly. Check it out:

$$-x - 3 = -2x + 1$$

Hmm, yeah . . . too many negatives for my taste. Multiplying both sides by -1, we get:

$$(-1)(-x - 3) = (-1)(-2x + 1)$$

Using the distributive property, we get $x + 3 = 2x - 1$.

Much better! Notice how each term's sign changed from negative to positive and from positive to negative. Remember, as long as we do the same thing to *each entire side* of the equation, we keep the scales balanced. If having negative variables unnerves you, definitely keep this shortcut as a tool in your back pocket!

Issue #4: I've Ended up with Too Many Parentheses!
Whether you're multiplying both sides by -1 or just changing subtraction $\to$ "adding negatives," here's my recommendation for when, suddenly, too many parentheses get involved: Use the distributive property, and distribute as much as you can as early as you can. For example, if you started out with $-3x - 1 = -2(x - 1) - 5(x - 2)$ and thought, "Hey, I don't like all these negative signs; I'll just multiply each entire side by -1 to neutralize them," you'd get:

$$(-1)[-3x - 1] = (-1)[(-2)(x - 1) - 5(x - 2)]$$

Um, YIKES! I mean, it would be SO easy to make a mistake. Instead, try distributing stuff *before* you multiply both sides by -1. So the right side of the equation, $-2(x - 1) - 5(x - 2)$, would become $-2x + 2 - 5x + 10$. Combining like terms, it becomes $-7x + 12$.

Now our original equation would look like this with no parentheses anymore: $-3x - 1 = -7x + 12$. At this point, with no parentheses, it's much easier to multiply both sides by -1:

$$(-1)(-3x - 1) = (-1)(-7x + 12)$$

Now we can distribute the -1 normally and with our sanity in check:

$$\to 3x + 1 = 7x - 12$$

And we could continue from there.

We're ready for more practice. If you run into any trouble, refer to the Workshop and the Troubleshooting Guide on the previous pages!

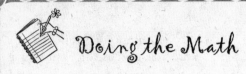

Doing the Math

Solve for *x*. Remember, we're finding the value of *x* that makes these statements true, following the step-by-step method on p. 181. I'll do the first one for you.

1. $5 - 2x + x = 3 - x$

<u>Working out the solution</u>: We need to collect all the *x*'s together on one side, so let's add *x* to both sides: $5 - 2x + x + x = 3 - x + x$, which becomes $5 - 2x + 2x = 3$. Further combining, we get $5 = 3$? Wait, what happened? The *x*'s disappeared, and we ended up with an *untrue* statement. This means it's never true, no matter what *x* is!

<u>Answer</u>: **No Solution**

2. $6x + 10 = 4(x + 3)$

3. $-2x - 5 = -x + 1$

4. $3x + 2 - x = -6 + 2x + 8$

5. $\frac{2x}{3} + 1 = x$ *(Hint: Multiply both entire sides by 3.)*

6. $x + 2xy + 1 - xy = 2x - 7 + xy$ *(Hint: Notice what happens to the xy term when you collect variables together and combine like terms correctly).*

(Answers on p. 321)

Takeaway Tips

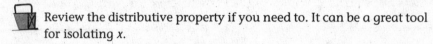When solving for *x*, think of unwrapping a gift. If you get confused about what order to do things in, remember that you are *undoing* PEMDAS.

Review the distributive property if you need to. It can be a great tool for isolating *x*.

The most important thing to keep in mind when solving equations is that you must <u>do the same thing to both sides of the equation</u>. That's how you keep the scales balanced, which is the only way to get a true answer in the end.

If there is more than one *x* term in the equation, always add and/ or subtract terms on both sides until all the *x* "stuff" is on one side and all the constants are on the other. Then combine like terms, and it's time to undo PEMDAS.

Always check your final answer by substituting it for *x*. If you get a true statement, you've found the right value for *x*!

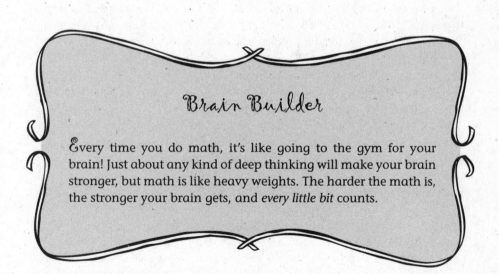

Brain Builder

Every time you do math, it's like going to the gym for your brain! Just about any kind of deep thinking will make your brain stronger, but math is like heavy weights. The harder the math is, the stronger your brain gets, and *every little bit* counts.

TESTIMONIAL

Pamela Fryman (Hollywood, CA)

<u>Before</u>: Self-conscious lunchtime loner
<u>Now</u>: Successful TV director and executive
producer!

In junior high, I never felt "cool." I always
seemed to be able to make people laugh, but even
still, I remember not
wanting to sit at the lunch
tables because I felt so
self-conscious. Instead, I
would wander around at
lunchtime, eating peanut
butter crackers in a corner
somewhere. I just never felt
comfortable.

"Leadership
positions are
natural for
women."

In math class, I worked hard and got B's. While I
wasn't at the top of my class, on some level, I
connected with math. I grew to love the logical
problem-solving strategies we learned in algebra. I
had no idea that, someday, this would help me in my
dream career—working in television comedy!

One of my first TV jobs was on a soap opera,
getting coffee and filing papers in a back office.
It wasn't much, but I was thrilled to be there. It
was after my first small promotion that I began to
use math: It was my job to keep track of the "time
code" (the digits updating the show's actual duration),
which was an exciting and ever-evolving puzzle! I
had to constantly add and subtract integers to
figure out how behind or ahead the show was running.
There was the target end time, the actors' varying
lengths of performances, and commercial breaks to
take into account. After a lot of hard work over a
few years, I was offered my first directing gig.

Today, I'm a director and executive producer of
television comedies like *How I Met Your Mother*, and
I use mathematical thinking more than ever. Logical
problem-solving skills are essential, especially

for "setting the shots": Devising strategies and coordinating four cameras at once to get the shots we need—in a small number of hours—is an ongoing challenge. Just like in word problems, I have to think very logically to solve each puzzle: "What do I have, and what do I want?" Every actor, every episode, and every show introduces new variables into the equation, so to speak, and with these real-life "word problems," there's never any time to waste!

As a woman in a male-dominated field, I'm pleased to say that I've experienced nothing but encouragement at every step of the way toward the leadership position I now hold—from women and men alike. If you think about it, leadership positions are natural for women. After all, my job involves managing not only logistical and creative puzzles, but also people. I suppose my teenage daughters have provided plenty of training in that regard! I love knowing that the crew is doing their best work—and not because I ever raise my voice, but because they feel respected and appreciated, and their talents nurtured. The result? Together we've created a hit TV show, and the set is a truly fun place to work. *(See p. 264 for Sudoku—the set's obsession!)*

Women are naturally nurturing and sensitive, and extremely capable. You can have as rich a future as you choose. I encourage you to challenge yourself in your math classes and to develop your logical problem-solving skills. No matter what career you pursue later on, these skills will train you to be not only well-rounded leaders, but well-rounded *women*.

Chapter 13

Nope, She Never Gets Off the Phone

Word Problems and Variable Substitution

$\mathcal{T}$o avoid getting ripped off, we gals need to be savvy when it comes to our phone bills. It's either that or learn how to spend less time on the phone.

Ha ha. That was a good one.

Let's say your best friend is spending a month abroad in Spain. I'm sorry, but there's no way you're not talking to her for a month, right? Every phone company handles long-distance calls differently, but let's say you had this choice:

Fabulous Fones: $14/month, plus $0.10/minute
Happy Talk: no monthly charge, but $0.30/minute

It seems like if you don't talk much, the Happy Talk company is the way to go. But how many minutes would you need to talk in order for Fabulous Fones to be the better deal?

In Chapter 11, we learned how to translate from English into math. Then in Chapter 12, we learned how to solve for x. Now we're ready to combine these talents and actually solve some serious word problems!

Let's first tackle the full version of some of those story problems we *translated* in Chapter 11. You'll find that once you've written the correct math translation for what's happening to the variable, solving the equation is *easy* by comparison.

Here's the full version of the problem on p. 152: "Connor wants to be taller. He said that if he multiplied his present height by 12 and then divided the result by 11, he'd be his ideal height. If his ideal height is 6 feet, what is Connor's present height?"

The first two sentences are the same as the problem on p. 152. We chose to label his present height p and his ideal height i. And just like before, we can read the problem and translate the relationship between them: $i = \frac{12p}{11}$.

Now, we're being told that his ideal height is 6. That means $i = 6$. It looks like by substituting 6 for i, we can rewrite our bigger equation: $6 = \frac{12p}{11}$. And now we can solve for p! To unwrap p, let's multiply both sides by 11 and we get $66 = 12p$. Next, let's divide both sides by 12, and we get $\frac{66}{12} = p$. Reducing the fraction, we get $p = \frac{11}{2}$; written as a mixed fraction, $p = 5\frac{1}{2}$ ft. = 5 ft., 6 in. (Remember: half of a foot equals 6 inches.)

Answer: Connor's present height is 5 feet, 6 inches.

With your new "solving for x" talents sharp, you're ready to fully solve the kinds of problems we *set up* in the DOING THE MATH section on pp. 154–6.

Doing the Math

Finish these word problems started on pp. 154–6. (Refer back to the answers you got, to save time.) I'll do the first one for you.

1.   Kim started off with a lot of bracelets. After buying 3 additional bracelets, she decided to share them all equally between herself and 9 friends for the day. If each friend got 2 bracelets, how many did Kim start off with? (Solve for s.)

<u>Working out the solution</u>: On p. 155, we discovered that the expression $\frac{s+3}{10}$, equals the number of bracelets each friend got, right? Now we're told that each friend got 2 bracelets, so that means $2 = \frac{s+3}{10}$. See what I mean? And now we can solve for s. First, we'll multiply both sides by 10 and get $(10)2 = \frac{(10)(s+3)}{10} \rightarrow 20 = s + 3$. Subtracting 3 from both sides we get $20 - 3 = s + 3 - 3 \rightarrow 17 = s$. And it's easy to check the answer. Let's see: Kim starts off with 17

bracelets, then buys 3 more (now she has 20), and then shares them equally with 10 people total. That's 2 each. Yep, we got it right!

<u>Answer:</u> Kim started off with 17 bracelets.

2.   After doubling the money in her bank account, Trudy spent $5 on music downloads. If she now has $195, how much money did she start off with? (Solve for *s*.)

3.   Brittany loves frozen red grapes. She has a whole bowl of them. She eats 5 frozen grapes and shares the rest equally among herself and her 5 friends: Anne, Nicole, Aliza, Paul, and Kirsten. Nicole got 7 grapes. How many did Brittany start off with? (Solve for *s*.)

4.   Chris had way too many text messages on his phone, and wanted to clear them out. He first deleted 10 text messages, but he still had way too many, so he deleted half of what was left. If he currently has 8 text messages in his phone, how many did he start out with? (Solve for *s*.)

5.   Sarah had a lot of ringtones in her phone yesterday. But she kept adding to them, and today she has 9 more than twice what she started out with. If she now has 51 ringtones, how many did she have yesterday? (Solve for *y*.)

6.   Suzanne works at a pet store. During the morning of her last day working there, she had lots of puppies sitting in the playpen. By the end of the day, she'd sold $\frac{4}{5}$ of them. Of those remaining, she took 2 home to keep for herself. And guess what? She didn't leave any puppies behind. How many did she start off with that morning? (Solve for *m*.)

(Answers on p. 321)

More Strategies for Word Problems

Sometimes the best way to start a problem is to make a chart or diagram and get a feeling for what's going on. Let's try solving the problem from the beginning of the chapter this way. Remember our goal: We want to know how many minutes we'd need to talk on the phone so that Fabulous Fones is the better deal, right? And here's the information we have:

Fabulous Fones: $14/month, plus $0.10/minute
Happy Talk: no monthly charge, but $0.30/minute

Hmm. We know that we're going to want to compare costs, eventually. So, translating, what value don't we know? We don't know the number of *minutes* we'll be talking on the phone. Let's label that m. Okay, now that we have a label m, let's write expressions for how much the total *cost* would be with each phone company. So, if I talked for m minutes, then:

Fabulous Fones would charge me: $14 + 0.10m$
Happy Talk would charge me: $0 + 0.30m$

Let's get a feel for what's going on by plugging in some values of m (5 min, 30 min, 2 hr, and 5 hrs) to see how much each company would charge:

# of minutes talked m	Fabulous Fones would charge $14 + 0.10m$	Happy Talk would charge $0.30m$
$m = 5$	$14 + (0.10)(5) =$ **$14.50**	$(0.30)(5) =$ **$1.50**
$m = 30$	$14 + (0.10)(30) =$ $17.00	$(0.30)(30) =$ $9.00
$m = 120$	$14 + (0.10)(120) =$ **$26.00**	$(0.30)(120) =$ **$36.00**
$m = 300$	$14 + (0.10)(300) =$ $44.00	$(0.30)(300) =$ $90.00

So, if you hardly talk at all, like if you talk for 5 minutes all month long (dream on!), then Happy Talk will be way less expensive: only $1.50 compared to $14.50! But, at a certain point, when m gets big enough, the two companies' costs will be equal to each other. Then, when m gets even bigger, Fabulous Fones looks much better.

For instance, if you talk for 5 hours (300 minutes) total during the month, then the HT company will charge you $90, and the FF company will charge only $44.

They seem to switch places somewhere between 30 minutes and 2 hours. So, how can we find the exact number of minutes after which the FF company becomes the better deal? We need to find *the number of minutes* that would result in both companies charging the *same* amount, right?

Look at what happens if we set these two "cost" expressions equal to each other:

$$\text{Cost at Happy Talk } = \text{ Cost at Fabulous Fones}$$
$$0.30m = 14 + 0.10m$$

By setting these two expressions equal to each other, we're saying "this cost over here IS equal to that cost over there." Yes, we're going to be that bold about it. By setting it up this way, we can now find the value of m that *does* make this statement true. I know it almost feels like backwards logic, but it works!

Let's solve for m:

$$0.30m = 14 + 0.10m$$
$$\rightarrow 0.30m - \mathbf{0.10m} = 14 + 0.10m - \mathbf{0.10m}$$
$$\rightarrow 0.20m = 14$$
$$\rightarrow \frac{0.20m}{0.20} = \frac{14}{0.20}$$

(Let's get rid of that decimal in the fraction!)

$$\rightarrow m = \frac{14 \times \mathbf{10}}{0.20 \times \mathbf{10}}$$
$$\rightarrow m = \frac{140}{2}$$
$$\rightarrow m = 70$$

So we've found the number of minutes that would result in each company charging us the *same* amount. But we saw from the chart that the longer we talk, the better deal the FF company becomes. So, now we can conclude that if you talk to your best friend in Spain for *longer* than 70 minutes, the Fabulous Fone company is the better deal. Only 1 hour and 10 minutes for the whole month? Yeah. I'd go with that company for sure.

By the way, this whole idea of "setting two things equal to each other" is a new concept—one you'd have no reason to think of unless you'd seen someone do it first. So don't worry if it never would have occurred to you!

Using Extra Labels—Why Not?

Word problems often ask a question, like "What will Lisa's age be in 3 years?" or "How much money did Kelly start out with?" When you see questions like this, you know what your *goal* is from the start. However, this isn't all it's cracked up to be. Sometimes knowing your goal can make things more confusing—you know, that head-spinning feeling of "how the heck am I supposed to figure *that* out?"

Yeah, we've all been there.

Some books will suggest that you start out by always labeling the thing you're looking for and not labeling anything else. Look, if you label the thing you're being asked to find and you immediately see how to solve it, then, by all means, go for it. But when it's confusing, I find it much more helpful to not worry so much about what you're *looking* for.

Instead, just start labeling things and gathering information. After you've labeled stuff, you can begin to write down *relationships* between the things you've labeled, and things will get much clearer.

For the phone minutes example we just did, we could have started by not only labeling m for minutes, we also could have used variables for how much the companies would *charge* you. For example, you could have said: Let's say c = the amount that the FF company will charge me during the month, so $c = 14 + 0.10m$, and k = the charge from the HT company, so $k = 0.30m$.

Then, after you made your chart and you realized you wanted to see when the two costs would be equal to each other ($c = k$), you'd set $14 + 0.10m = 0.30m$.

You'll find that you drop some of the extra labels pretty quickly, but extra labels have always helped me to organize my thoughts during challenging word problems. So go ahead, get label-happy!

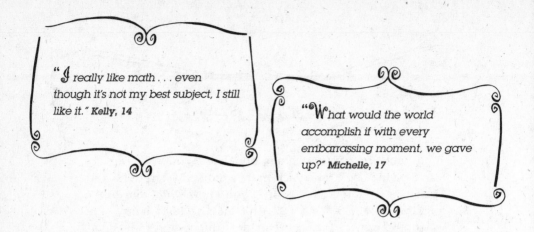

Variable Substitution

This is a technique that comes in really handy when solving word problems, especially when you're using more than one variable to label things. Basically, it comes down to substituting one variable for another variable expression.

To solve word problems, our goal is always to write down an equation using a *single variable*, and then to use the techniques we learned in Chapter 12 to *solve* the equation. But as we've talked about, sometimes it's easiest to first translate the story problem using two or more variables, and sometimes this results in two or more equations. But then what? Let's see how variable substitution can help!

Before I show you how to apply this technique to solving word problems, let's take a look at the nuts and bolts of this type of variable substitution.

Let's say we were told that $y = 2x$ and $3x + 5y = 26$. Hm. We can't solve either of these equations on their own; they both have two variables in them. But from these two equations, we can create a new equation that uses only *one* variable! Here's how:

The first equation, $y = 2x$, tells us that whatever x is, y will have twice that value. So, in the second equation, we can just substitute $2x$ wherever we see y. Using $y = \mathbf{2x}$:

$$3x + 5y = 26$$
$$\rightarrow 3x + 5(\mathbf{2x}) = 26$$
$$\rightarrow 3x + 10x = 26$$

Now that we have an equation with a single variable, we can solve it, no problem:

$$\rightarrow 13x = 26$$
$$\rightarrow x = 2$$

And now we can easily find y by plugging $x = 2$ into the first equation, $y = 2x$:

$$y = 2x$$
$$\rightarrow y = 2(2)$$
$$\rightarrow y = 4$$

If you plug $x = 2$ and $y = 4$ into the original equations, you'll get two true statements!

Next, we'll apply this concept to word problems. But first, here's the step-by-step method to guide us through the process.

Step By Step

Setting up and solving word problems when you don't know how to start!
(This might seem long, but it can be used to solve tons of word problems. So read it a few times, read the examples a few times, and you'll be in great shape!)

Step 1. Look for the values you *don't know*, and label them with variables of your choice. If you can write everything in terms of one variable, that's best. But you can also use more than one variable if it's easier.

Step 2. Read the problem again and figure out *what happens* to the variables (for some types of problems, drawing a picture or chart at this stage helps). Now translate this into math, writing down everything you know in math language.

Step 3. Look for how to *establish relationships between values*, and set up the indicated equation(s). "Is"-type verbs will offer a good clue for where to place an =. And if it helps to set two expressions equal to each other (like in our phone example on pp. 195–6), consider doing that!

Step 4. If you've decided to use more than one variable, now that everything's organized, it's time to figure out how to write an equation in terms of just *one* variable. This usually involves *variable substitution*: Substitute shorter equations into bigger equations.

Step 5. Solve the equation. Double-check to make sure you've answered the question that was asked and that the answer makes sense.

There will be many word problems that you'll read carefully and realize, "Hey, I think I see how to do this." But some will be just baffling. When you don't have a clue as to how to solve one of these problems, the above STEP BY STEP can help!

 *And... Action!*

Step By Step In Action

Here's the full version of the Annie problem on pp. 151–2:

"Annie's age is doubled and then subtracted from 21. The result is equal to Danny's age. Danny is the same age as Annie. What is Annie's age?"

Step 1. Huh? Okay, we don't panic. We simply start labeling and trust that more will become clear. Let's label Annie's age: Let Annie's age = a. Reading the problem, I see another unknown value, Danny's age. So, let Danny's age = d. Also, notice that the first "is" does not imply an equals sign. Just pay attention, and you'll be able to tell the difference.

Step 2. What happens to the variables? Well, a is doubled and then subtracted from 21. As we saw on p. 151, that's $21 - 2a$. Nothing "happens" to d, it seems.

Step 3. Let's look for the "is"-type verbs for establishing relationships between values. After Annie's age is doubled and subtracted from 21, it "is" equal to Danny's age, so $d = 21 - 2a$. Any other relationships? Well, it says that Danny's age *is* her age. That means $d = a$, right? We've got two equations with two variables!

Step 4. Now it's time to write our equation with only *one* variable, using variable substitution. Our short, easy equation is $d = a$, and since $d = 21 - 2a$, we can substitute a for d and write: $a = 21 - 2a$.

Step 5. Now we solve the equation $a = 21 - 2a$. To get all the a's on one side, let's add **2a** to both sides and get:

$$a + 2a = 21 - 2a + 2a \rightarrow 3a = 21$$

Divide both sides by 3 and get $\frac{3a}{3} = \frac{21}{3} \rightarrow a = 7$. And that's what the problem asked for!

Let's see if our answer makes sense: If Annie is 7, and we double her age ($7 \times 2 = 14$), then subtract it from 21, we get $21 - 14 = 7$. Yep, we got her age back again!

Answer: Annie is 7 years old.

 QUICK NOTE   Remember: Your goal is to get a single equation, written in terms of a single variable, which you'll then be able to solve.

Let's do a more challenging one. We'll end up with three equations and three variables!

 *Take Two: Another Example*

Amanda, Davana, and Emily all have the same phone, and they've all started collecting ringtones for them. Amanda has twice as many ringtones as Davana, and Emily has 3 more ringtones than Amanda. If, together, they have a total of 103 ringtones, how many ringtones does Davana have?

Step 1. Don't panic: Just label and organize! I see three unknowns. Let **a** = # of Amanda's ringtones, **d** = # of Davana's ringtones, and **e** = # of Emily's ringtones. So far, so good?

Step 2. In this problem, nothing really "happens" to the variables, so let's move on.

Step 3. Look for relationships between values. Okay, it says that Amanda has twice as many ringtones as Davana. That means **a = 2d**. (You can always plug in values to see if you put the 2 in the right spot; that's what I always do.) Next it says that Emily has 3 more ringtones than Amanda: **e = a + 3**. (We could also write this as $e = 2d + 3$. Do you see why?) Also, it says that together they have 103 ringtones. That means **a + e + d = 103**, right? Reread the problem. Did we miss any information? Nope, so let's move on. And lookie there, we're totally organized, aren't we?

a = 2d

e = a + 3

a + e + d = 103

Step 4. Write an equation in terms of just one variable. Hmm, how do we do this? Well, we can look at just the first two, easier equations: **a = 2d**, and $e = a + 3$. Then we can substitute $2d$ for a, and the second equation becomes **e = 2d + 3**. Does that make sense? Now we can look at the longer equation, and wherever we see a we can write $2d$, and wherever we see e, we can write $2d + 3$:

$$a + e + d = 103$$

Substituted!

$$(2d) + (2d + 3) + d = 103$$

Notice how each variable, a, e, and d, is expressed in terms of the d variable.

Step 5. Now let's solve the algebra problem. First, bring all the d's together.

$$2d + 2d + d + 3 = 103$$
$$\rightarrow 5d + 3 = 103$$
$$\rightarrow 5d + 3 - 3 = 103 - 3$$

$$\rightarrow 5d = 100$$
$$\rightarrow \frac{5d}{5} = \frac{100}{5}$$
$$\rightarrow d = 20$$

So Davana has 20 ringtones. And that's what the question asked for! Notice that from this information, we also could figure out how many ringtones Emily and Amanda have. From the equation $a = 2d$, we know Amanda has $2(20) = 40$ ringtones. In other words, $a = 40$. And from the equation $e = a + 3$, we know that Emily has $40 + 3 = 43$ ringtones. In other words, $e = 43$.

Checking our answers: $20 + 40 + 43 = 103$. Yep!

Answer: Davana has 20 ringtones.

Phew! That one was trickier, but I wanted you to see the power of using a few short equations to create a single equation with *one variable*, which you could then solve to get the answer.

QUICK NOTE   When you have no idea how to start a story problem, remember: Just start labeling stuff, and looking for relationships between things. You may be surprised at what unfolds before your eyes!

Doing the Math

Solve these word problems using the step-by-step method above and variable substitution where necessary. I'll do the first one for you.

1. Darcy has 9 fewer songs in her playlist than double the number Sarah has. Together, they have 81 songs. How many songs does Darcy have?

<u>Working out the solution:</u> Start by labeling! Let d = # of Darcy's songs, and let s = # of Sarah's songs. We know that $d + s = 81$, right? Also, translating that first sentence into math, we get: $d = 2s - 9$. We have two equations and two variables, so let's use variable substitution and put **2s − 9** where we see d in the first underlined equation. Substituting for d, we get **(2s − 9)** $+ s = 81$. Dropping the parentheses, combining like terms, and adding 9 to both sides, we get $3s = 90$, so $s = 30$. But the problem asked for the number of Darcy's songs. Let's plug $s = 30$ into our second equation, $d = 2s - 9$, and get $d = 2(30) - 9 \rightarrow d = 51$. And it's easy to check in our first equation that $30 + 51 = 81$. Yay!

<u>Answer:</u> **Darcy has 51 songs.**

2.   A purse and matching shoes cost a total of $95. If the shoes cost $55 more than the purse, how much does the purse cost? *(Hint: Don't jump to conclusions here!)*

3.   Your mom has offered you $4/hour to clean the garage, plus a $10 bonus if you start by 8 A.M. on Saturday. Your dad has offered you $4.50/hour to clean the basement. Each job would take all weekend. Before you decide which offer you'll accept, you want to figure out this: **a.** Assuming you start by 8 A.M. on Saturday, if you only want to work for 15 hours this weekend, which offer is better?
b. How many hours would you have to work for their offers to be equal? *(Hint: This is like the phone example on page 195.)*

4.   Hunter is 3 years older than Duncan. Leslie is 4 years younger than Duncan. Together, their ages equal 41. How old is Leslie? How old is Duncan? How old is Hunter?

(Answers on p. 322)

Takeaway Tips

When confronted with a word problem that you don't know how to start, begin by labeling things. Then find relationships between the things you've labeled, and translate English into math.

If you end up with more than one equation and more than one variable, use substitution to create just *one* equation with just *one* variable—and then solve for it!

Keep your brain on, and always reread the problem after you get your answer. Sometimes the variable you actually solve for isn't the value the problem asked you to find.

Danica's Diary

REALITY VS. IMAGINATION

Imagination is a wonderful thing. It lets us dream big, and it can drive our ambitions to make those dreams come true. Yet imagination can also work against us if we're not careful. If you read the Danica's Diary on pp. 157-8, you saw how I made all sorts of assumptions about what that guy calling me *meant*. In my mind, I was walking down the aisle with him and choosing our children's names—and I mean, gimme a break, I was still in junior high! I guess if we really get our heart set on a guy, it can be easy to imagine things that just don't exist. People say that "seeing is believing," but sometimes "believing is seeing." I think you know what I mean.

Reading into things happens in more than just love; it happens in *all* parts of life. It's easy to jump to conclusions or think we see something that isn't there. Just like in #2 on p. 204. Did you think the answer would be $40? Most people do when they first see it. The key in math is to be very literal when you're reading a problem: What is *actually* there on paper? What information does it *actually* give you? Then, after translating it very carefully into math, you'll see that the right equation usually presents itself to you. THEN you'll understand how the problem is going to go, but not necessarily before that point. It's normal to start the problem not knowing how it's going to resolve itself. And the more realistic and literal you are about the translating, the less often you'll get that dizzy, "lost" feeling. It's hard to be dizzy when both feet are solidly on the ground and you're dealing with exactly what is in front of you, right?

Some of this can apply to life. I think there's an advantage to being able to see what is actually in front of us, nothing more and nothing less, just like in word problems. But look, I'm an actress, too, so I know very well that imagination is wonderful and so emotionally thrilling!

Some people decide they're just not going to imagine or "hope" for anything, because that way they won't be disappointed. I don't happen to agree with that philosophy. I think the goal in life must be to let our imaginations play, and let our dreams fill us with all sorts of wonderful feelings, but then not to get disappointed when any one *particular* thing doesn't happen. After all, there could be so many wonderful surprises (careers, guys, etc.) that we might not have imagined . . . and that *reality* just might deliver!

? ? ? ? ? ? ? ? ? ? ? ? ? ? ? ? ? ?

Lateral Thinking Puzzles

Sometimes we *have* to make assumptions based on common sense. Like if you kiss your best friend's boyfriend, you can *assume* that you'll lose a friend . . . and self-respect. Then again, just because you've never liked zucchini doesn't mean you should assume you won't love zucchini bread. They sound alike, but they're completely different. In fact, as you may know, zucchini bread doesn't taste anything like zucchini; it's more of a dessert!

Learning to recognize which things we *know* and which things we are *assuming* (even reasonable assumptions) is a powerful skill to develop—both in math and in life. Here are a few "assumption" puzzles to exercise that recognition skill.

There are exactly five strawberries in a small basket.
Five people each take one of the strawberries.
How can it be that one strawberry is left in the basket?

(Don't read further if you want to figure this one out on your own!)

Solution: After reading this question, most people assume that the last person took the strawberry *out* of the basket, right? That's the assumption that keeps us from seeing an easy solution: When there was just one strawberry left in the basket, the last person could have taken the last strawberry by taking the whole basket!

How about this one? It's sort of a classic; it's been around forever.

A man and his son are in a car crash.
The father is killed, and the child is taken to the
hospital, badly injured.
When he gets there, the surgeon says,
"I can't operate on this boy. He is my son!"
How can this possibly be?

(Hint: There's no priest involved or anything else of that nature. If you *really* need a hint on this one, read near the bottom of p. 164. And then you can feel very ashamed of yourself!)

Here's a good one:

A puppy fell out of a 20-story building—and lived! How is this possible?

(Hint: It's a normal, huge, 20-story building; the puppy didn't have a parachute; and yes, it's a normal puppy that does not know how to fly. It also doesn't matter what the puppy landed on. And, for that matter, it has nothing to do with luck.) Promise me you'll think about this for at least 30 seconds before looking at the answer! For the answer, see the footnote below.*

Notice how the *images* we create in our mind "fill in" information that we haven't been told. We assumed things about the strawberries, the surgeon, and the puppy. Then those assumptions show up in the images in our minds, limiting our ability to tell the *difference between* our assumptions and the actual information we were given. This happens during math homework, too, especially with word problems.

There are other kinds of lateral thinking puzzles like these visual ones:

$$\frac{man}{board} \rightarrow \text{"man overboard"}$$

$$|r|e|a|d|i|n|g| \rightarrow \text{"reading between the lines"}$$

Try this one on your own: $\frac{stand}{i}$

Notice how solving these visual puzzles also requires being able to see what is "literally" in front of you. You can do an Internet search for "kids lateral thinking puzzles" and find many

. . . .

* The puppy fell out of the first-floor window.

more of these puzzles; in fact, there are whole books full of them, if you really get into it.

Becoming truly aware of our *assumptions* can make us better solution finders to all sorts of puzzles and problems in math—and in life!

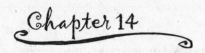

Can a Guy Be *Too* Cute?
Intro to Solving and Graphing Inequalities

*T*oo cute? You might think I'm crazy for even asking such a question. Your motto might be "The cuter the better—bring it on!"

Alright then, on a cuteness scale from 1 to 10, with 1 being your scary Halloween mask and 10 being Zac Efron or Antonio Banderas, where do you want your guy to be? If we assign the variable *c* to his cuteness, then *c* would need to be, say, 7.5 or higher? And just for fun, let's say there are even values beyond 10. In fact, since you don't think there's such a *thing* as too cute, let's go ahead and include *all* values higher than 7.5, and including 7.5:

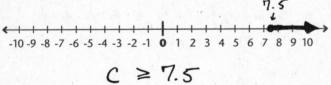

$$c \geq 7.5$$

Any of those values for *c* will satisfy you, right? But looking at that ray* we've drawn, we graphed numbers like 11.5, 163, 1 million, and beyond!

I'm not sure I'd want to date a guy who was *that* pretty. Hmm, let's rethink this. I mean, some of these guys probably can't help but check their hair in every mirror they walk by. You know the type. They pretend not to know that they're flexing their arm muscles every 5 minutes, but it's totally deliberate?†

. . . .

* A **ray** is a line that has an endpoint on one side and an arrow on the other, indicating that it goes on forever.
† It's not necessarily their fault. Some of them were probably raised to believe their self-worth comes from their appearance alone, so that's all they think about.

Also, from my experience, the more you get to know and like a guy, the cuter he becomes—I mean, *way* cuter. So let's keep the mobs of superficial girls away from our guys and get smarter with this graph. Personally, I'm thinking about keeping the *c* value somewhere in the 5 to 8.75 neighborhood.

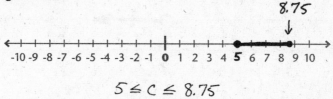

$$5 \leq c \leq 8.75$$

Ah, much better. Staying on this totally superficial thread, how tall do you want him to be? Let's say that *h* = the *difference* between his height and your height. In other words: If 0 means you're the same height, then 3 would mean he's 3 inches taller than you, and −3 would mean he's 3 inches shorter than you. Got it? So, what kinds of *h* values would you prefer? (Remember: If he's too tall, he's out of spontaneous kissing range!)

How about these graphs:

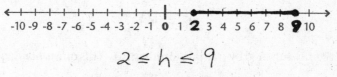

$$2 \leq h \leq 9$$

(If you always wear tall shoes)

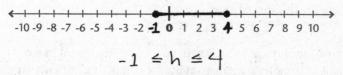

$$-1 \leq h \leq 4$$

(You know, to stay within kissing distance. Plus, short guys are adorable, and lots of them just haven't finished growing yet.)

Some girls insist that the guy has to be taller, but it doesn't matter how *much* taller, so any value higher than 0 would work. This means we don't want to include 0 on our graph, just *every number* above it. By the way, when I say "every number" I'm including all rational and irrational numbers.* Yep, we'll be including crazy numbers like 0.000001 inches, $\sqrt{2}$ inches, $\frac{100}{3}$ inches, and even π inches.†

To graph "all numbers bigger than zero" (and *not* including zero), we just draw an empty circle at 0 and continue the arrow from there.

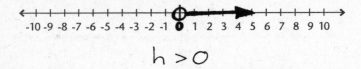

$$h > 0$$

Remember, the more values you allow on each graph, the more options for guys you have. (Some math problems are satisfied by a whole bunch of values, too. These are called *inequalities*, which we'll learn to solve in a few pages.)

We could also do this exercise with much more important attributes—the ones that make good boyfriend/husband material, like kindness, intelligence, loyalty, and so on.

Of course, on a scale from 1 to 10 of how much a guy *respects* you:

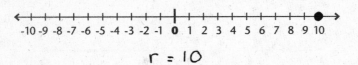

$$r = 10$$

. . . that's only satisfied by a single number: 10. (This one is *non*negotiable, ladies!)

Graphing Inequalities

I want to emphasize that when you see something like $x < 5$, this statement is a math sentence saying "x is less than 5," which indicates an *infinite* number of values. Every single value smaller than 5 will satisfy that inequality: 4.9, 0, -6.314, etc. So its graph looks like this, with an endpoint of $x = 5$ and the ray extending forever in the negative direction.

· · · · · · · · · ·

* For definitions of *rational* and *irrational numbers*, see pp. 313–4 in the Appendix.
† The value $h = \pi$ would mean he's approximately 3.14 inches taller than you.

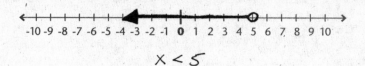

$$x < 5$$

I like to call the value $x = 5$ the "endpoint," because it's at one end of the graphed solutions ray. Makes sense, right?*

Notice how we didn't fill in the circle at the endpoint $x = 5$, because we don't want to include it in the solution. The $<$ (is less than) symbol means that we want every value *less* than 5, but not 5 itself. We only fill in the circle when we see a $\leq$ (is less than *or equal to*) or $\geq$ (is greater than *or equal to*) symbol. This way, we *include* the endpoint as part of the solution. Easy, right?

QUICK NOTE   In an expression of inequality, the variable can appear on either side of the symbol. For example, $1 < x$ is the same as $x > 1$, and their graphs look identical. (I still like to think of the "is greater than" symbol an as alligator mouth, wanting to eat the bigger value.) It's easier to interpret it when the variable is on the left, but both math sentences are saying the same thing: "1 is less than x" and "x is greater than 1" mean the same thing. In fact, I recommend saying them out loud, especially when the variable is on the right side, just to help the concept sink in a bit more!

By the way, if you haven't done so recently, check out the formal definition of **inequality**, along with a full review of the symbols, on pp. 146–8.

· · · · · · · · · ·

* Many algebra textbooks call this type of value a *critical value*. But I prefer the term *endpoint*. I'm guessing you prefer it, too!

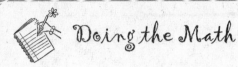 *Doing the Math*

Graph these inequalities. I'll do the first one for you.

1. $-4 < y \le 5$

<u>Working out the solution:</u>

Notice that I used an empty circle for the endpoint -4, because the $<$ means we can't include it, and I used a filled-in circle for the endpoint 5. Also, notice that even though this inequality's solution set has two endpoints, there are still an infinite number of values! *

2. $6 > w$

3. $-9 \ge x$

4. $2 < n < 3$

(Answers on p. 322)

Solving Inequalities

Remember how we graphed the different values for guys' cuteness and height on pp. 210–2? Each time, a whole bunch of values satisfied our preferences. Understanding what it means to "solve an inequality" all

• • • • • • • • • •

* In fact, there are an infinite number of points inside the segment from 0 to 1. See p. 315 for more on this!

comes down to realizing that some math sentences can be satisfied by a whole bunch of values, too.

When we solve equations like $2x - 7 = 1$, we find the *one* value of x that makes the equation true. In this case, only $x = 4$ satisfies the requirements of the equation, and its graph is easy:

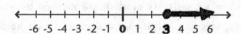

So far, we've been solving equations. But how about solving *inequalities*?

$$2x + 1 \geq 7$$

For a moment, let's translate the math statement above: "Twice x, added to 1, is greater than or equal to 7." Which values of x make this sentence true?

Actually, the method is practically the same as the one for solving equations! Just do the same thing to both sides of the equation, with the goal of isolating x:

$$2x + 1 \geq 7$$
$$\rightarrow 2x + 1 - \mathbf{1} \geq 7 - \mathbf{1}$$
$$\rightarrow 2x \geq 6$$
$$\rightarrow \frac{2x}{\mathbf{2}} \geq \frac{6}{\mathbf{2}}$$
$$\rightarrow x \geq 3$$

Notice that because of the $\geq$ symbol, the endpoint ($x = 3$) *is* included in the graph, so we use a filled-in dot. And what have we done? We've found the entire **solution set**—all the values of x that satisfy the inequality $2x + 1 \geq 7$. And notice from the graph that the solution set actually has an *infinite* number of values.

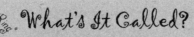

What's It Called?

Solution Set

A **solution set** is the group of *all* of the numbers that make a math statement true.

Examples:

The *solution set* for the equation $x + 1 = 3$ is simply $x = 2$.

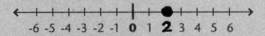

The *solution set* for the inequality $x + 1 > 3$ is *all* numbers greater than 2, in other words: $x > 2$.

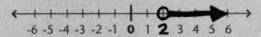

You'll learn formal notation for solution sets later in algebra. For now, as long as you understand that inequalities can have a whole bunch of solutions, you'll be fine!

QUICK (REMINDER) NOTE   Back in Chapter 3, we learned that when we multiply or divide a number by –1, we get its opposite; it's like holding up a mirror! If we hold up a second mirror, we get the opposite of the opposite, which is the original thing back again. For example: $(-1)(-1)(5) = 5$. You'll see why I'm reminding you of this in a moment.

Multiplying and Dividing Negatives in Inequalities

This is probably the most important (and strangest) rule for solving inequalities:

The Mirror Rule for Inequality Symbols

When we *multiply* or *divide* both sides of an inequality by a negative number, we must REVERSE the direction of the inequality symbol.

So, any time we multiply or divide both sides of an inequality by a negative number, we must not only change the signs of all the terms, but also change the *direction* of the inequality symbol! For example, $<$ becomes $>$, $\geq$ becomes $\leq$, and so on.

If this seems strange to you, think about this: If multiplying and dividing by -1 is like holding up a mirror, then what do you suppose a $>$ symbol looks like in the mirror, hmm?

That's right: When we hold up a mirror to each side of the inequality (in other words, when we multiply each side of an inequality by a negative number), the inequality symbol is affected by the "mirror," too! Let's think about numbers for a moment.

You know that $-4 > -5$, right? (If you're blanking, just think about where they each live on the number line: -4 is to the right of -5, so it's bigger.) Well, multiply both sides of the inequality $-4 > -5$ by -1, and look at what you get if you *don't* reverse the inequality sign: $4 > 5$. Yikes!

Here's another true statement: $-5 > -10$. Look what happens if we divide both sides by -5 and *don't* reverse the symbol. We get $1 > 2$. Yikes again!

This is something that just doesn't come up with equals signs. If we have something like $-x = -5$, then we know we can multiply both sides by -1 and get our answer: $x = 5$. I guess that's because $=$ looks the exact *same* in a mirror.

So, if we have an inequality like $-x < -5$, for example, we have to multiply both sides by -1 *and* reverse the symbol to get the final answer: $x > 5$.

The Mirror Rule for reversing inequality symbols (see p. 217) only works for multiplication and division, not addition or subtraction. You can safely add and subtract negative numbers from both sides of an inequality and be just fine, so don't reverse the direction of the symbol unnecessarily!

Solving Inequalities with the Mirror Rule:

Let's see why the rule works by solving the inequality $10 - x \geq 4$. Just eyeballing it, we notice that if $x = 6$, the two sides will be equal, right? And if x is even *smaller* than 6, say $x = 3$, then the statement is still satisfied, because $10 - 3 \geq 4$. So, I'm going to guess that if x is 6 or *less*, this statement is going to be true. That would be this solution set: $x \leq 6$. Now let's see how the solving goes:

$$10 - x \geq 4$$
$$\rightarrow 10 - \mathbf{10} - x \geq 4 - \mathbf{10}$$
$$\rightarrow -x \geq -6$$

Using the rule we just learned, to get x completely isolated now, we need to multiply both sides by -1 *and* reverse the direction of the inequality:

$$\rightarrow -x \geq -6$$

$\downarrow$ reverse it!

$$\rightarrow (\mathbf{-1})(-x) \leq (\mathbf{-1})(-6)$$
$$\rightarrow x \leq 6$$

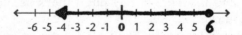

And that's the answer we predicted! Notice that we can pick any random value in the solution set, like $x = \mathbf{0}$, plug it into the original inequality, and we'll get a true statement:

$$10 - x \geq 4 \rightarrow 10 - (\mathbf{0}) \geq 4 \rightarrow 10 \geq 4$$

Yep, true!

QUICK NOTE   When doing homework, it's important to always <u>check your answer</u>. For inequalities, we do this by first checking the *endpoint* we found. Then we plug a random value from the solution set into the original inequality and make sure we get a true statement. This second step confirms that we got the correct *direction of the symbol*.

Step By Step

Solving and graphing inequalities:

Step 1. Do things to both sides of the inequality until x is isolated on one side, just like in solving equations. Each time you multiply or divide both sides by a *negative* number, however, use the "Mirror Rule" and *reverse* the direction of the inequality symbol. The answer will be an entire solution set, and the number you see will be the endpoint on the graph.

Step 2. Now graph the solution set on a number line. If the inequality uses $<$ or $>$, use an empty dot at the endpoint. If the inequality uses $\leq$ or $\geq$, use a filled-in dot at the endpoint.

Step 3. Just like when we solve equations, check your answer by plugging the endpoint into the original inequality. After simplifying, if you end up with *the same number on both sides of the inequality symbol*, that means you got the correct endpoint.

Step 4. Now pick an "easy" value from the solution set (you can look on your graph to pick one) and plug it into the original inequality to check the direction of the symbol. If you get a true statement, that means you've found the correct direction of the symbol!

Watch Out!

In Step 3, when you check your endpoint by plugging it into the original inequality, if the symbol is $<$ or $>$, then you'll actually end up with an untrue statement like "$4 < 4$." That's okay. You haven't made a mistake! The only important thing is that you get the same number on both sides. If you do, then you've found the correct <u>endpoint</u>.

QUICK NOTE   In Step 4, we pick a random value from the solution set and plug it into the original inequality to check our answer. I always check to see if 0 is in the solution set. If it is, then I plug that value in, to keep the math easy. Other easy values to plug into equations tend to be $x = 1$, $x = -1$, $x = 10$, and $x = -10$. If you've done it right, you'll end up with a true statement like "$5 < 6$." I'll show you how this works now!

Step By Step In Action

Find and graph the solution set for:

$$3 - 2x \leq 5$$

Step 1. Isolating x, we'll first subtract 3 from both sides:

$$3 - \mathbf{3} - 2x \leq 5 - \mathbf{3}$$
$$\rightarrow -2x \leq 2$$

Let's divide both sides by -2, and this means we'll need to use the "Mirror" rule from p. 217 and *reverse* the direction of the symbol from $\leq$ to $\geq$.

$$\frac{-2x}{-2} \geq \frac{2}{-2}$$
$$\rightarrow x \geq -1$$

Step 2. Graphing the solution set, $x \geq -1$, we'll be sure to use a filled-in dot at the endpoint $x = -1$, because the symbol means we're including the -1:

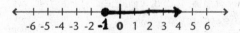

Steps 3 and 4. To check the answer, we plug the endpoint $x = -1$ into the *original* inequality:

$$3 - 2(-1) \leq 5$$

$$\rightarrow 3 + 2 \leq 5$$

$$\rightarrow 5 \leq 5$$

We got the same number on both sides, so we have the correct *endpoint*, $x = -1$. But just because we found the correct endpoint, this doesn't mean we didn't make a mistake with the direction of the inequality symbol. So let's check that, too: We'll pick a value within the solution set to make sure our ray is pointing in the correct direction. According to our graph, we can use anything greater than -1. We're lucky; $x = 0$ is in the solution set! That makes it easy. Plugging 0 into the original inequality:

$$3 - 2(\mathbf{0}) \overset{?}{\leq} 5$$

$$\rightarrow 3 \leq 5 \checkmark$$

Yep! We got a true statement, which means we got the entire solution set correct!

Answer: $x \geq -1$

Take Two: Another Example

Solve and graph the solution set for:

$$-7 - x < -12$$

Step 1. Let's do things to both sides of the inequality to isolate x. But you know what? Because there are so many darn negative signs, let's start out by *multiplying each entire side by* -1; of course, this means we'll also need to reverse the direction of the symbol:

$$-7 - x < -12$$

↓ *reverse it!*

$$\rightarrow (-\mathbf{1})(-7 - x) > (-\mathbf{1})(-12)$$
$$\rightarrow 7 + x > 12$$

Ah, much better. Now we just subtract 7 from both sides and get $x > 5$.

Step 2. Graphing it, we'll be sure to use an empty circle at $x = 5$, so we *don't* include the endpoint (see below for graph).

Step 3. Checking our endpoint, we substitute 5 for x and get $-7 - (\mathbf{5}) < -12 \rightarrow -12 < -12$. Now this obviously isn't a true statement, because $<$ means "is strictly less than and *not* equal to." But that doesn't matter; we were just making sure we got the right endpoint, and you can see that we did!

Step 4. Checking the solution set, what's an easy value satisfying $x > 5$ that we can use to substitute into $-7 - x > -12$? Hmm, we can use anything bigger than 5. Unfortunately, we can't use 0; that's my favorite one to substitute. Let's use 10. So, we get $-7 - (\mathbf{10}) < -12 \rightarrow -17 < -12$. Yep, we got a true statement!

And now we know we've gotten the right solution set!

Answer: $x > 5$

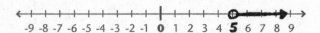

 *Doing the Math*

Find and graph the solution set for these inequalities. In other words, what group of values of *x* make these *true* statements? I'll do the first one for you.

1. $-3(x - 9) > 6$

<u>Working out the solution:</u> Alright, to isolate x, let's first divide both sides by −3. Doing this means we'll need to reverse the symbol, so we get $\frac{-3(x - 9)}{-3} < \frac{6}{-3}$. On the left, the two −3's cancel, and on the right, the fraction simplifies* to $\frac{6}{-3} = (-1)\left(\frac{6}{3}\right) = -2$. So, we now have $(x - 9) < -2$. We can safely drop the parentheses and then add 9 to both sides: $x - 9 + 9 < -2 + 9 \rightarrow x < 7$. And there's our solution set!

Let's check our answer: To check the endpoint, we can plug x = 7 into the original inequality and simplify: $-3(7 - 9) > 6 \rightarrow -3(-2) \rightarrow 6 > 6$. The two numbers are the same, so we know the endpoint is correct. And because 0 is in the solution set x < 7 (it's smaller than 7, after all), we can plug x = 0 into the original inequality and make sure we get a true statement: $-3(0 - 9) > 6 \rightarrow 27 > 6$. Yep! Our whole solution set checks out

<u>Answer:</u> x < 7

~~~
 ←|—+—+—◀—+—+—+—+—+—+—+—◯—+—+—→
  -9 -8 -7 -6 -5 -4 -3 -2 -1  0  1  2  3  4  5  6 **7** 8  9
~~~

2. $8 + x < 16$

3. $8 - x < 16$

. . . .

* For more on dealing with negative fractions, see p. 46.

4. $-3x - 1 \geq 5$

5. $2x - 1 > x + 3$ (Hint: Start by subtracting x from both sides!)

(Answers on p. 322)

What's the Deal?

What's up with the "Mirror Rule" for multiplication/division by negative numbers? Why does it work . . . *really*? Well, believe it or not, the Mirror Rule is actually like a *shortcut* for adding stuff to both sides: Anytime we multiply or divide by a negative number, we could use addition or subtraction to just swap the two sides' places instead.

Take a look at this: $-x \geq -6$. (Try writing these yourself, too.)

Method: Just adding stuff to both sides	Method: Using the Mirror Rule
$-x \geq -6$	$-x \geq -6$
→ $-x + 6 \geq -6 + 6$ (Add 6 to both sides.) → $-x + 6 \geq 0$	(We'll multiply both sides by (-1) and *reverse* the direction of the symbol.)
→ $-x + x + 6 \geq 0 + x$ (Add x to both sides.)	→ $(-1)(-x) \leq (-1)(-6)$
→ $6 \geq x$	→ $x \leq 6$

And remember, $6 \geq x$ is the same exact statement as $x \leq 6$. Well, how about that!

Takeaway Tips

 If we multiply or divide both sides of an inequality by a negative number, we must also use the Mirror Rule and *reverse the direction of the inequality symbol.*

 Aside from that Mirror Rule, solving inequalities is *just* like solving equations: We do the same thing to both sides, until we get *x* all by itself on one side of the inequality. Once we've isolated *x*, we are now looking at the answer—the solution set!

The solution sets for inequalities can be graphed using rays, indicating an infinite number of values that all satisfy the inequality.

Remember that < or > means we graph an open circle at the endpoint, and ≤ or ≥ means we use a closed circle.

You Said: Your Horror Stories About Procrastination!

Do you put things off until the last minute and then wish you hadn't? Are you trying to improve from bad past experiences? Read these stories—you're not alone!

"Last week, instead of doing my homework, I got on MySpace for a long time. I knew it was getting late, but I just really wanted to stay on the computer. When I finally got off, it was already dark—it was after 8:00! I still had chores to do, and I had to eat. There was no way I could do my homework. I learned that I should never get on MySpace before I do my homework." **Daisy, 13**

"Junior year, I had to write a term paper for American history. I put it off for weeks until my mom made me pick a topic and get my sources, but I still didn't start writing. The Sunday night before it was due, I was getting butterflies in my stomach because I didn't think I was going to finish, and it would be the first time I had ever turned something in late. Well, I finally finished it at 3 A.M. that night, but as I clicked print . . . the computer froze. I panicked, but there was nothing I could do! Why do computers always freeze when you're running late? When I woke up a few hours later, it was still frozen so I had to go to school without it. Lesson learned!" **Sarah, 17**

"Our big term paper was assigned, and I knew for weeks that my research notes were due by a certain date, but of course I kept putting things off—until 2 days before! I went to the library to get books, only to find that many books that would have been helpful were already checked out. The next night I had to do 50 note cards, and they had to be from 10 different sources. I was up 'til two in the morning and didn't even finish. I learned from this experience a little, but not that much. I've gotten better about big assignments, but I still put off homework until the last minute. I don't know why I do it. I need to stop!" **Amy, 16**

"Surprisingly, I enjoy my trigonometry class, and my teacher is amazing. The thing is, though, my teacher doesn't collect homework until the end of the semester . . . so I kind of stopped doing it. I always intended to 'catch up' the next night or on the weekend, but hockey games and basketball games kept coming up, and trig got put on the back burner. But, now that the semester is coming to an end, I realize how far behind I actually am! From now on, I'm going to try to do my homework when it is supposed to be done, whether it is 'due' or not." **Stacey, 17**

"'Don't worry, Mom; I'll work on it Sunday.' This would usually be my response to my mother whenever she would tell me to work on my homework or college searching, or even scholarship essays. As soon as the deadline would come around, I'd hurriedly scribble down some words. My chances of receiving a scholarship from the lame, last-minute essay were slim to none. Procrastination is a terrible habit to get into. No matter how much you would rather be having fun, whenever something important needs to be done, it should be your top priority. You can always have fun later. When you finish your project, all your stress is relieved. Accomplishment is a great feeling." **Nathan, 17**

"My first term paper was almost impossible to write, and I had to pull some long nights right before it was due, since I kept putting it off. The day it needed to be turned in, I was going on a trip and wouldn't be at school, and I still had to proofread it and write the bibliography. Fortunately, the woman who was driving me, Emily, was bringing her laptop. I figured that since it was a six-hour drive, I could write during the drive. But she didn't bring her battery! I ended up turning in the paper late and got a C. I found that you can't put off a six- to ten-page paper until the last week, you can't expect to always have time to work on things, and you can't expect people to do exactly what you think they'll do, to 'save' you when you're running behind schedule." **Mary, 16**

"Our class had to do a big report, but I waited until three days before it was due to come up with a topic: optical illusions. I worked really hard those three nights, and I even brought in posters to explain things better. I thought I did a good job, but I ended up getting a really bad grade. I guess the real 'illusion' was the idea that I might be able to do a good job in only three days!" **Adrienne, 12**

"Last year, I had to write my first term paper. I worked really hard on it, and I got a great grade. The next semester, I had to write another term paper—this time, I let my good grade go to my head and procrastinated. Even when I realized that I was running out of time, I didn't work on the paper as much as I should. When I eventually turned it in, I knew it wasn't very good. The whole time that I was waiting to get it back, I was stressing about what a horrible grade I was going to get. And when I finally got my paper back, sure enough, I'd gotten a rotten grade. This year, I had to write another term paper, and I made a conscious decision to work on it a little bit every day. I did, and I got a good grade! It's funny how that works." **Jessica, 18**

Poll: What Guys Really Think . . .
About Talented Girls

We asked more than 200 guys, ages 13 to 18 (in anonymous, multiple-choice surveys) what they think about girls who succeed in sports and school. Here's what they had to say!

If you knew a girl did better than you in a sport, would you be embarrassed?

GUYS SAY:

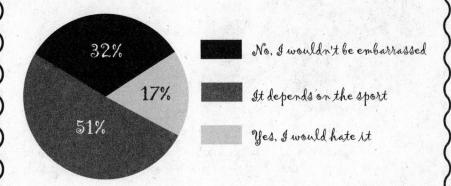

32%

17%

51%

■ No, I wouldn't be embarrassed

■ It depends on the sport

■ Yes, I would hate it

More than half of the polled guys have a pretty open mind; they said it would depend on the sport. After all, some of them might think it would be normal for a girl to beat them at gymnastics or tennis, but not at something like wrestling or rugby.

Almost a full $\frac{1}{3}$ of guys (remember, $\frac{1}{3} \approx 33.3\%$)* would not be embarrassed at all, no matter what the sport is! I think that's pretty impressive. Maybe they didn't think of the rugby thing.

Finally, 17% (that's almost $\frac{1}{5}$, since $\frac{1}{5} = 20\%$) admit that they'd be embarrassed and would hate for a girl to be better than them at sports—any sport.

. . . .

* That's an approximation, which is why I used the $\approx$ symbol. To be precise, $\frac{1}{3} = 33.\overline{3}\%$. The bar over the 3 means that the 3's go on forever! To review repeating decimals, check out Chapter 10 in *Math Doesn't Suck*.

Let's be real: Physiologically speaking, most guys have more muscle mass than most girls, especially in high school and beyond. So guys really do have a bit of an edge in sports that require brute strength. But remember that sometimes being fast, flexible, and agile is just as important as being physically strong. So get out there, and show 'em what you got!

If you knew a girl did better than you on a test,
would you be embarrassed?

GUYS SAY:

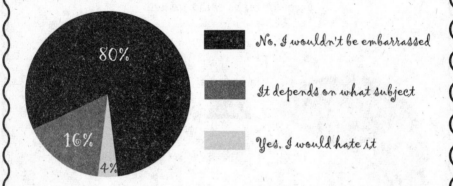

80%

16%

4%

No, I wouldn't be embarrassed

It depends on what subject

Yes, I would hate it

This one speaks for itself! Guys might feel competitive in sports, but according to this poll, most guys seem much more comfortable with girls beating them on tests.

Some girls pretend to get bad grades just so guys won't feel embarrassed. But according to this poll, guys aren't as fragile as you might think. Selling yourself short by pretending to get bad grades only hurts you in the long run—you'll lose people's respect. Of course, nobody likes a show-off (not girls *or* guys), so even though it's important to not hide your smart and savvy side, you also don't want to parade around with your A+ test in hand!

Hmm . . . not showing off, but not hiding your smarts, either? It's a delicate balance to strike, but one that's well worth striving for. You'll use it not only to deal with test grades now, *but also with things you'll accomplish in your successful career someday!*

For more guys' polls, see p. 75!

Quiz:
Do You Pick <u>Truly</u> Supportive Friends?

*A*re your friends looking out for your best interests? What kinds of people do you choose to spend time with?

Take this quiz by expert psychologist Dr. Robyn Landow and contributor Anne Lowney, and see how you fare!

1. You're at one of those up-all-night sleepovers when conversations turn from gossip and boys to more serious things: doubts, fears, hopes—all that kinda stuff. When you start to bring up your future as a high school student, a college student, or even an adult in general, how do your friends react?

a. They hear what you're saying, but it seems like they don't really relate. Maybe they haven't thought that much about life after middle school, but at least they're *listening*.

b. They feel relieved to hear you have the same questions as they do about the future. You all know that although there's tons of fun ahead, there will be important choices to make in the coming years, too.

c. They tell you it's way too early to start worrying about all that stuff and wonder why you bring up school (of all things!) when you're supposed to be having fun.

2. You've noticed that your math class has gotten a lot harder lately. To tell the truth, most of the material is flying right over your head! Fortunately, the girl in the seat behind you always knows what's going on—and she seems willing to help! Only thing is, she's not part of your group of friends; in fact, you've never really talked to her before. You:

a. Go ahead and ask her to explain a few difficult equations, knowing that she can help you out a lot. Plus, you can pass on some of her tricks to your friends who are struggling as well. And maybe you'll even make a new friend!

b. Avoid her because you know your friends consider her a "nerd"— one of those girls who only cares about school. They would definitely make fun of you for talking to her, and it's just not worth it.

c. Hesitate. Your friends might be threatened by you making friends with an "outsider," but they'd probably understand that you need help, right? So, you decide to ask for her help only after you've exhausted all other options, like struggling on your own and asking the teacher.

3. All day you've been complaining to your friends about the mountain of work you need to finish for your biology assignment. As you're searching the web and asking yourself how there could possibly be *this* much information about fruit flies out there, your friends keep trying to chat with you. When you tell your friends to cool it with the IMs because you need to focus, what's their next message?

a. "u 4 real need to chill . . . fine, cya."

b. "k, talk 2 u later."

c. "What r u talking about? Everyone's on and I gotta tell u what's up with . . ."

4. You're really excited about your last test grade in science. You aced it! When your friend asks you how you did, you:

a. Act like you bombed the test just to satisfy a friend who does not care about school, much less a stupid science test.

b. Proudly tell her you got an A+! After all, you studied all weekend. You deserve it.

c. Say you did okay—same old story. You don't want to seem like you're bragging.

5. How would you describe your friends' attitudes toward your teachers?

a. If they aren't talking or laughing during class, they're passing notes behind the teacher's back. You end up getting in trouble because of them, even if you didn't do anything! It's annoying, but what can you do? They're your friends, after all.

b. The teachers are pretty much considered the "enemy," but your friends know how to keep it to themselves—at least until class is over and you can all agree that your history teacher's pit stains are *way* out of control.

c. Of course, there are always one or two teachers they secretly can't stand, but from time to time, your friends will admit to admiring a teacher for something—like for his or her intelligence or patience.

6. You've got a big history test coming up, and you really don't want to face it alone. When you suggest a study group among friends, how do they respond?

    a. One or two of your friends decide to study with you, but the others seem disinterested.

    b. They can't understand why you make such a big deal over tests and stuff. Wouldn't you rather be hanging out and relaxing rather than getting all stressed over school?

    c. They're all for it! Why not tackle the beast together? You can help each other out and provide some much-needed moral support.

7. We all know parents can be way overprotective. Still, you've got to admit, every once in a while they get it right: Maybe they foresee a problem before it actually happens, sense when you are upset, or just give some really good advice. So, when it comes to your friends, what are your parents' thoughts?

    a. They always remark on what nice friends you have. They know them well and like to stay caught up on their lives. They realize they are smart, respectful, and, most of all, a good influence on you. Overall, there's a very positive vibe going.

    b. They don't really know your friends that well. You're reluctant to talk about them too much because you're afraid your parents might be judgmental or disapprove of them in some way—maybe their clothes, their looks, or the way they talk or act.

    c. Your parents have plainly said that they think your friends may not be a good influence on you. They are always encouraging you to dump your friends and meet new people.

8. It's finally the night of your school orchestra concert. After all the practice you've done over the past few months, you are so excited to perform and show off what you have accomplished. You've got a couple of family members coming. Are your friends just as supportive?

    a. When you invited them, they were like, "Yeah . . . cool." They show up to your performances about half the time you invite them, so you figure there's a 50/50 chance they'll show up.

    b. Of course! They know how important music is to you and how much work goes into preparing for a performance. Their smiling faces are the first thing you see when the curtain comes up!

    c. Not exactly. The few times you've mentioned your music, they haven't seemed that interested. They're your friends, of course, but you don't talk about that kind of stuff with them; it's like in a different category of your life.

9. Every year, your Spanish teacher picks one or two students who really stand out to take the National Spanish Exam for middle school students. You're surprised and flattered when she recommends *you*, noting that you have one of the highest grades in the class. She warns you that it will require a bit more study outside of class, and that the test lasts four hours—so you have to be willing to make the commitment. You're so proud of yourself for being chosen that you decide to sign up. When you tell you friends, how do they respond?

    a. "Why would you want to spend all that time studying for a test that isn't even part of our school grade? You've got to be crazy! You already think way too much about school. Why add to the work you have to do?"

    b. "That's awesome that the teacher recommended you. You must be doing really well in Spanish. You should definitely take the test! I hear if you get the award, it's announced at our end-of-the-year ceremony."

    c. "That's cool that she brought it up with you. But, do you really like Spanish that much that you'd want to do more studying?"

10. You have your first boyfriend, and spending all your time with him has been amazing! But after two weeks of this, your schoolwork is really starting to suffer. You know you need to strike a better balance in the future, but for now, you're totally behind. When you tell Mr. Wonderful that you can't see him this weekend because you're bogged down with homework, how does he take it?

    a. He agrees but doesn't understand why you need the *whole* weekend to do your work. He'd rather you guys hang out at least once, maybe see a movie.

    b. He's bummed, but he completely understands. Truthfully, he's gotten behind on his schoolwork, too, and he admires your discipline.

    c. He makes a big deal out of it and gets all insecure that you don't want to spend time with him. You're his girlfriend! The truth is, he can be pretty possessive.

Scoring

1. a = 2; b = 1; c = 3	6. a = 2; b = 3; c = 1
2. a = 1; b = 3; c = 2	7. a = 1; b = 2; c = 3
3. a = 2; b = 1; c = 3	8. a = 2; b = 1; c = 3
4. a = 3; b = 1; c = 2	9. a = 3; b = 1; c = 2
5. a = 3; b = 2; c = 1	10. a = 2; b = 1; c = 3

If you scored between 10–14

Keep up the good work with your choice of friends! You pick people to spend time with because they're caring, supportive, and totally there for you; they care about school; and of course, they're fun to be around! And it shows, because they've got your back, and they help you feel better about yourself. As you know, a true friend knows when to hang around, when to give you space, when to listen to what is bugging you, and when to cheer you up so you can end a bad day on a high note.

Make sure that you are taking the time to be a good friend, too. Some people don't always know how to ask for support when they need it. Sometimes they need to be prompted! With good friends by your side, life can be so much better, and you can inspire each other to live up to your full potential!

If you scored between 15–24

Your friends might get distracted sometimes, but generally speaking, they're *there* for you when you need them. It sounds like you've chosen a diverse group of people, too, which is great. You've got your study partners, your confidantes, and your "fun" friends. As you know, different friends can be good for different purposes. Of course, it's up to you to figure out how to maintain that balance!

Pay attention to which friends are your *real* friends. These are the good listeners, supporters, and overall cheerleaders of your success and happiness; be sure to hold onto them through thick and thin. Remember, it's easy to find people with great connections or who are popular, but it's much harder to find someone who is a true friend with the same priorities as you.

Also, since you're in the middle of the scoring scale, I recommend reading both other sections. You'll get something from each of them!

If you scored between 25–30

Hmm. Listen to your intuition when you're around your "friends." Some people give off positive energy that makes us feel good, while others give off negative energy that drains us. Who's truly happy when you succeed? Who's understanding when you need time to yourself or to study? Get to know these people better. You may find that your shopping pal is also an excellent study buddy! Who knew?

On the other hand, who makes regular *demands* on your time and loyalty? What may seem like loyalty or closeness could actually be possessiveness. If you pay attention, you'll know the difference.

Don't worry, you don't have to "fire" your friends; you'll have opportunities to be with them in group social settings. But who do you want shaping your life? After all—and this is true—we *become like* the people we surround ourselves with.

If you aren't totally satisfied with what you find in your current circle, be open to new people! It's easy—get involved. Whether it's a hobby you love or a charity you want to volunteer for, you can find opportunities to meet people with the same values and interests as you. You never know when that new friend will come into your life and what unique gifts she will bring. Make a point of strengthening your relationships with those who encourage and inspire you. Most of all, *be* a real friend, and you'll *attract* that kind of friend.

Chapter 15

Champagne and Caviar

Intro to Exponents

*H*ave you ever dreamed of being a high-powered executive? You know—corner office on the top floor, the interns getting you coffee. Being so powerful, you wouldn't have to *do* very much to get big results. Say, for example, you wanted to set up a company-wide meeting or even throw a big office party. You could just make a phone call and ask someone to plan it all for you.

Imagine: You stroll into your office, casually toss aside your designer briefcase, throw your feet (in fabulous shoes) up on your polished mahogany desk, and dial your assistant. *Ring!*

Sexy guy's voice: "Good morning, what can I do for you?"

You: "Hello Brett, I'd like to throw a little shindig here at the office on Friday afternoon. Would you make that happen? You know—caviar, champagne, the works. Send invites to everyone. They've all been working so hard lately. They deserve it."

Then, poof! Come Friday, with no more effort on your part, *voilà!* There's a party, complete with your favorite champagne and caviar. That must be how *exponents* feel.

Let me explain.

Exponents

So, what are exponents anyway? Here's what an **exponential expression** looks like:

$$4^3 \quad \leftarrow \text{exponent}$$
$$\nearrow$$
$$\text{base}$$

The 4 is called the *base*. This is easy to remember, because it's like the base of a statue—the bottom part that holds everything up. The *exponent* is the little number up top, in her corner office on the top floor. She's really powerful. Here's why: Exponents are actually shorthand for *multiplication*.

Say you want to multiply $4 \times 4 \times 4$. You could write it out, or you could just write 4^3. They mean the same thing. The high-powered executive "Ms. Exponent" is so powerful that all she has to say is "3," and suddenly the 4 is multiplying times itself 3 times!

$$4^3 = \underbrace{4 \times 4 \times 4}_{3 \text{ of them}}$$

Here's another example:

$$2^7 = \underbrace{2 \times 2 \times 2 \times 2 \times 2 \times 2 \times 2}_{7 \text{ of them}}$$

As you can see, an exponent is a powerful little number "upstairs," which represents *repeated multiplication*.

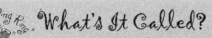

What's It Called?

Exponential Expression

This is just a fancy way of saying "an expression with at least one *exponent* in it." Some examples of exponential expressions are 3^4, $6 \cdot 2^3$, and even $4xy^5$.

Exponent

An *exponent* is the little number in an exponential expression that indicates <u>how many times</u> the base gets multiplied times itself. For example, in the expression 3^4, the exponent is **4**, so the 3 should multiply times itself **4** times:

$$3^4 = 3 \times 3 \times 3 \times 3 = 81$$

Base

Like the base of a statue, the *base* is the bottom number in an exponential expression. It's the thing that's going to be *doing* the multiplying, according to the instructions from Ms. Exponent. For example, in the exponential expression 3^4, 3 is the base, and it will multiply times itself 4 times.*

How Do I Say It?

There are a few different ways to say these types of expressions aloud. Here are the most common ones—the ones you're most likely to hear your teacher say in class. Notice that the 2 and 3 exponents have special names; this is because they represent the physical dimensions we use for area and volume (like for a square and a cube).

3^2 x^2	Three squared x squared
3^3 x^3	Three cubed x cubed
3^4 x^4	Three to the fourth x to the fourth
3^5 x^5	Three to the fifth x to the fifth

(The higher numbers are all just like these last two.)

. . . .

* Bases can be variables, too, which we'll see in Chapter 16.

QUICK NOTE   Numbers can also have exponents of 1. In fact, *most numbers* have an exponent of 1. You don't have to write it because it's kind of silly: $5 = 5^1$. However, later in algebra, knowing this tidbit will come in handy, so it's a good thing to know!

Watch Out!

Some people might see 3^3 and think that the answer is 9, because they see the two 3's and just do multiplication. But as you know, $3^3 = 3 \times 3 \times 3 = 27$. Or they might think 7^3 equals 21—yikes! Remember, exponents are WAY more powerful than regular multiplication. So, when you see an exponent sitting up there in her corner office, remember how powerful she is and write out what the expression *means* if it helps. Go ahead and write $7^3 = 7 \times 7 \times 7$, and avoid making mistakes that come from underestimating the power of Ms. Exponent.

QUICK NOTE   The great thing about exponents with a base of 10 is that, in these cases, the exponent tells you exactly how many zeros there will be! So, $10^1 = 10$, $10^2 = 100$, $10^3 = 1,000$, and so on.

Doing the Math

Using exponents, write these expressions in much shorter, more efficient ways. Don't worry about *evaluating* them; just *rewrite* them. I'll do the first one for you.

1. $6 \times 8 \times 6 \times 6 \times 8$

<u>Working out the solution</u>: First, let's separate the 6's from the 8's. We count three 6's and two 8's: $(6 \times 6 \times 6) \times (8 \times 8)$, and now it's much easier to see the answer.

<u>Answer:</u> $6^3 \times 8^2$

2. $2 \times 2 \times 2 \times 2 \times 2 \times 5 \times 5$

$2^5 \times 5^2$

3. $10 \times 10 \times 10 \times 10 \times 10 \times 10$

10^6

4. $12 \times 12 \times 7 \times 7 \times 12 \times 7 \times 12$

$12^4 \times 7^3$

5. $(0.2) \times (0.2) \times (0.2) \times (0.2)$ *(Hint: Treat (0.2) like any other number, and leave the parentheses in place.)*

(0.2^4)

(Answers on p. 322)

QUICK NOTE   If your base is the number 1, no matter what exponent you use, you'll always end up with just 1. This makes sense: $1^2 = 1 \times 1 = 1$, and $1^3 = 1 \times 1 \times 1 = 1$, etc.

So, written as a universal rule, $1^m = 1$, for all real numbers m. Again, we're not making headlines here, but it's good to keep in mind!

Danica's Diary

A "POWER"-FUL KITTEN!

I don't usually do "email forwards," but every now and then I'll get the cutest little video of a baby kitten chasing her tail or something, and it just captures my heart! The looks on their eager little faces are so cute, and I suddenly want other kitty lovers to feel as happy as I do.

This happened last night, and I immediately sent the kitten video to 5 friends and family members. Awash in happy kitten thoughts, I began to wonder how many people, total, might see the video as a result of my action.

Let's say that my friends and family send it to 5 people each, and then those people send it to 5 *more* people each, and so on. If it takes a full day for each person to respond (and assuming that each time the video is emailed to 5 *new* people), then how many people total will have seen the kitty video by the end of the week?

Hmm. Let's do this one step at a time—the best way to tackle ALL math problems.

I sent the video to 5 people. So, at the end of the first day, 5 *new* people will have seen it. So far, so good.

At the end of the second day, if all 5 people send it to 5 people *each*, that would be 25 *new* people, right? Think about this for a second. Hold out your hand and look at it. Pretend that your fingers are the 5 people I first sent the video to. Now, if each of those people send the video to 5 new people, imagine a whole new little hand sprouting out of the end of each finger! Those 25 *little fingers* would represent all of the *new* people who see it on the *second* day.

Now, on the third day, those 25 people each send the video to 5 new people. So, imagine each of those 25 *little fingers* sprouts another teeny-tiny hand. How many new teeny-tiny *fingers* are on those 25 hands?

That's $25 \times 5 = 125$. This represents the number of *new* people who would see the kitten video on the third day.

1st day: $5 = 5^1$ *new* people see the video
2nd day: $25 = 5^2$ *new* people see the video
3rd day: $125 = 5^3$ *new* people see the video

See the pattern?

Now can you tell me how many *new* people would see the kitten video on the seventh day? Yep, it would be 5^7.

So, the total number of people who will have seen the video after a week (not counting me) would be:

$$5^1 + 5^2 + 5^3 + 5^4 + 5^5 + 5^6 + 5^7 = ?$$

Now that we've done the hard part, it's time for a calculator!

$$5 + 25 + 125 + 625 + 3,125 + 15,625 + 78,125$$
$$= \mathbf{97,655}$$

And if you count me, that's 97,656 people.

Wow! That's nearly 100,000 people! Well, it really was a cute kitty.

There's a good chance that not everyone forwarded the email when they received it, but some may have sent it to 10 people or more! I like thinking about how many people I might have made feel happy in a week, *and I only had to send it to 5 people.* Can you figure out how many more days it would take to reach 2 million people? (It's less time than you think.)

Who Is Ms. Exponent Touching?

Exponents only affect their own base; they only "talk to" the base they're touching. For example, in the following expression, the exponent 3 only applies to 2, the base it's *touching*: $6 \cdot 2^3 = 6 \times 2 \times 2 \times 2 = 48$.

This probably goes without saying, but if you see something like $(-3)^4(6)$, the exponent is *not* considered to be "touching" the (6). The exponent would have no effect on the (6); it's on the wrong side of it!

Also, if the exponent is touching the outside of a parenthesis, the exponent will affect *everything inside the parentheses*: $(3 \times 2)^3 = 6^3 = 216$.

Knowing this tidbit will also help us handle negative number bases more easily.

Exponents and Negative Numbers

I would imagine that one of the challenges of being a high-powered executive would be managing all of the different types of personalities of the people working for you. What if you had some employees who were great at their job, but their attitude was sort of . . . *negative*? Like, maybe they had huge (and easily bruised) egos, or they thought that they were always right? Dealing with these *negative* types would require being more careful . . . And so it is for exponents.

If you were asked to evaluate 3^4, you'd know what to do: $3 \cdot 3 \cdot 3 \cdot 3 = 81$. But what about $(-3)^4$ or (-3^4)? And by the way, *these have two different answers*! Notice the difference? Look carefully at the placement of the exponent.

Remember, the exponent doesn't care *what* is inside the parentheses it's touching. It will multiply "it" times "itself" 4 times, no matter what "it" is. So, regardless of whether the value inside the parentheses is 3 or -3, the exponent will treat it the same.

$$(-3)^4 = (-3) \cdot (-3) \cdot (-3) \cdot (-3)$$

We end up with four negative signs, and we know from integer multiplication (see p. 43) that an *even* number of negative signs means they all cancel each other out!

So $(-3) \cdot (-3) \cdot (-3) \cdot (-3) = 3 \cdot 3 \cdot 3 \cdot 3 = 81$. In other words, $(-3)^4 = 81$, a positive answer.

However, what happens when the exponent is *inside* the parentheses?

$$(-3^4) = ?$$

Notice that we can drop the parentheses, because the whole job of parentheses is to separate things from each other. In this case, the parentheses are *not* acting as a shield or barrier between numbers or exponents, or anything else for that matter. So we can confidently say:

$$(-3^4) = -3^4$$

Look at this for a second, and make sure it makes sense to you; notice that in no way have we changed the value of anything.

Now, remember that the exponent only cares about exactly what it's *touching*: If the exponent is touching a set of parentheses, it includes *everything* inside the parentheses. If the exponent is only touching a number, it only affects that one number, and it doesn't care about anything else.

$$-3^4 = -(3) \cdot (3) \cdot (3) \cdot (3) = -81$$

Here's another way to think about it: On p. 41 we saw how we can rewrite a negative sign as "multiplication by (-1)," so $-3^4 = (-1)3^4$, right?

And as you know, this simplifies to: $(-1)81 = -81$, a negative answer. So, remember:

$$(-3)^4 = 81$$
$$(-3^4) = -81$$

Because this is *exactly* the kind of thing that teachers use to try to stump us on tests!

What's the Deal?

If you're wondering why you can't "pull out" a negative sign from the first example, $(-3)^4$, and then end up with a negative answer, read on: You *can* separate the negative sign from the 3, but <u>not</u> from the *parentheses*; it would look like this:

$$(-3)^4 = (-1 \cdot 3)^4$$

That didn't help things! You can't move the negative sign outside the parentheses, because it's *inside* a set of parentheses that an exponent is *touching*. Ms. Exponent already knows about the negative sign, and she expects it to obey her commands just as much as the 3 does.

Until you are super comfortable with this stuff, I really recommend writing out *what the exponential expression means* by writing out the multiplication like I've been doing here. Then you know exactly what you're dealing with! It'll save you from making TONS of errors with these negative signs and other stuff. Trust me, I've been there.

High-powered executives can surely deal with negative numbers. As you can see, it just requires a little more care and attention along the way!

Step By Step

Evaluating exponential expressions with negative signs *in them:*

Step 1. For each base with an exponent: What's the exponent actually touching? Pay attention to *exactly* where the negative sign is, if there is one. Does the exponent *affect* the negative sign? In other words, is the exponent *touching* parentheses that have the negative sign *in them*?

Step 2. If the negative sign is *not* affected by the exponent, then evaluate the expression as usual (by multiplying things out), and leave the negative sign where it is.

Step 3. If the negative sign *is* affected by the exponent, then evaluate the expression as usual (by multiplying things out), and:

• If the exponent is odd, the answer will be negative.

• If the exponent is even, the answer will be positive.

Step 4. Do this for each term (base) that has an exponent. Done!

Let's evaluate $(-2)^3 - 4^2$.

Let's start by looking at the first term, $(-2)^3$.

Step 1. Yep, the negative sign *will* be affected by the exponent, because it's inside parentheses that the exponent is touching. (We can skip **Step 2.**)

Step 3. The exponent is odd, so we'll end up with a negative answer. Let's multiply it out: $(-2)^3 = (-2)(-2)(-2) = -8$.

Step 4. Okay, next term: -4^2.

Now we can either think of this as subtraction, or *as adding a negative number*, but whichever way we think of it, the negative sign will *not* be affected by the exponent. So Steps 1 and 2 tell us we can just multiply it out, $4^2 = 16$, and leave the negative sign where it is:

$$(-2)^3 - 4^2 = -8 - 16 = -24$$

Done!

Answer: $(-2)^3 - 4^2 = \mathbf{-24}$

QUICK NOTE   From Your Little Friends, 1 and –1.
If your base number is (–1), then for any whole number exponent, you'll always end up with an answer of either 1 or –1, depending on whether the exponent is odd or even. So $(-1)^m = 1$ for all even m, and $(-1)^n = (-1)$ for all odd n. For example: $(-1)^2 = 1$, and $(-1)^3 = (-1)$.

 *Doing the Math*

Evaluate the following exponential expressions. I'll do the first one for you.

1. $-2^6 + (-9)^2$

Working out the solution: First, let's look at -2^6. What does that negative sign mean? It's stuck on the outside; the exponent doesn't even know it's there. Thinking about it another way, we can pull out the "factor" of -1, because the negative sign is not inside parentheses: $-2^6 = (-1)2^6 = (-1)64 = -64$.

Now let's tackle the $(-9)^2$. The negative sign is *inside* the parentheses that the exponent is touching, so the exponent *does* affect the negative sign. Since the exponent is *even*, the negative signs will cancel each other away, and we'll get a positive answer: $(-9)^2 = (-9)(-9) = 81$. Now we can rewrite our original problem like this: $-2^6 + (-9)^2 = -64 + 81 = 17$.

Answer: $-2^6 + (-9)^2 = 17$

2. $(-5^2) = ?$

3. $-5^3 - (-5)^2 = ?$

4. $-2^6 - 9^2 = ?$ *(Hint: This will have a different answer from #1.)*

5. $(-184.5)^4 - (184.5)^4 = ?$ *(Hint: You do NOT need a calculator to solve this one. Don't even think about it, missy. Start by writing it out in expanded form, but don't multiply anything out. Look at the negative signs, and see what happens!)*

(Answers on p. 322)

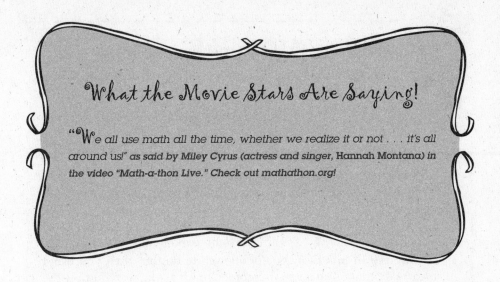

Exponents with Decimals and Fractions

As you may have noticed in problem #5 of the DOING THE MATH section on p. 248, the base number in an exponential expression can be a decimal, or heck, even a fraction. After all, why shouldn't this be true?

$$(0.095)^4 = (0.095) \times (0.095) \times (0.095) \times (0.095)$$

Or this?

$$\left(-\frac{1}{3}\right)^3 = \left(-\frac{1}{3}\right) \times \left(-\frac{1}{3}\right) \times \left(-\frac{1}{3}\right)$$

Well, both *are* totally true—thanks for asking! In fact, it doesn't matter *what* the base number is, if there's an exponent touching it, it needs to multiply *times itself* however many times Ms. Exponent tells it to.

Shortcut Alert: **Exponents and Fractions**

For any fraction $\frac{a}{b}$, look at what we can do with an exponent m:

$$\left(\frac{a}{b}\right)^m = \frac{a^m}{b^m}$$

We can "pull in" the exponent and stick it on the top *and* bottom of the fraction.

Let me explain: By now you know that $\left(\frac{2}{3}\right)^3 = \frac{2}{3} \times \frac{2}{3} \times \frac{2}{3}$, and good ol' fraction multiplication* tells us to multiply across the top and bottom, so this equals $\frac{2 \times 2 \times 2}{3 \times 3 \times 3} = \frac{8}{27}$.

But instead of writing out the fractions, this shortcut allows us to get the same result faster, by "pulling in" the exponent: $\left(\frac{2}{3}\right)^3 = \frac{2^3}{3^3} = \frac{8}{27}$.

If you keep in mind *why* you're allowed to do this, then you won't misuse the shortcut. And if you ever forget how it works, just multiply the whole thing out. *When in doubt, multiply it out!*

Exponents with Absolute Values

This is pretty straightforward but definitely worth mentioning. You know that $(-2)^3 = -8$, but what do you think $|-2|^3$ equals? Hmm, well: The expression $|-2|^3$ means "take the distance from -2 to zero, and multiply that distance times itself, 3 times." And the distance from -2 to zero is positive 2. So $|-2|^3 = (2)^3 = 8$. Makes sense, right?†

On the other hand, you might see something like $-|2|^3$, which you can see $= -8$, because the negative sign is *outside* the absolute value bars. In fact, $-|2|^2 = -(2)^2 = -4$.

. . . .

* To review fraction multiplication, see p. 51 of *Math Doesn't Suck*.
† Check out Chapter 4 for a review of absolute values . . . and spa treatments.

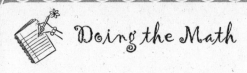
Doing the Math

Evaluate these expressions using fractions, decimals, and absolute values with exponents. Pay close attention, and you'll be fine! I'll do the first one for you.

1. $4\left|-\frac{3}{2}\right|^3 = ?$

<u>Working out the solution:</u> Any time we evaluate expressions, we should use PEMDAS* as our guide. Absolute value bars take the same priority as <u>P</u>arentheses, but there's nothing inside them that needs to be simplified. So next we do the <u>E</u>xponent. What's the exponent touching here? It's touching $\left|-\frac{3}{2}\right|$, which means "the distance from $-\frac{3}{2}$ to zero" needs to get multiplied times itself, 3 times. Distance is always positive, so we can drop the negative sign and replace the absolute value bars with parentheses. Now the problem looks like this: $4\left(\frac{3}{2}\right)^3$. Another way to get to this stage is to "when in doubt, multiply it out" and rewrite the entire expression as $4 \times \left|-\frac{3}{2}\right| \times \left|-\frac{3}{2}\right| \times \left|-\frac{3}{2}\right|$. Knowing that each $\left|-\frac{3}{2}\right| = \frac{3}{2}$, then we can rewrite this again as

$4 \times \frac{3}{2} \times \frac{3}{2} \times \frac{3}{2} = 4\left(\frac{3}{2}\right)^3$. See? Next we can either do a bunch of fraction multiplication, or we can use our shortcut from p. 250 to distribute the exponent to the top and bottom of the fraction, and the problem now looks like this:
$4\left(\frac{3^3}{2^3}\right) = 4\left(\frac{27}{8}\right)$. (The exponent didn't affect the 4 because the 4 was *outside* the parentheses that the exponent was touching.) Now that we're done with the exponents, we can <u>M</u>ultiply the 4, and we get $4 \times \frac{27}{8} = \frac{4}{1} \times \frac{27}{8} = \frac{\cancel{4}^1}{1} \times \frac{27}{\cancel{8}_2} = \frac{27}{2}$.

<u>Answer:</u> $4\left|-\frac{3}{2}\right|^3 = \frac{27}{2}$

. . . .

* See p. 21 for a review of PEMDAS. ("E" stands for exponents.)

2. $\left(\frac{3}{4}\right)^2 - \frac{11}{16} = ?$

3. $(0.5)^2 - |-0.25| = ?$

4. $-27\left|-\frac{2}{3}\right|^3 = ?$ *(Hint: This is very similar to the worked example above.)*

5. $-3|(-1)^2 - (-1)^3|^2 = ?$ *(Hint: Start by simplifying just what's inside the bars.)*

(Answers on p. 322)

QUICK NOTE Parentheses Are Your Friends
Use 'em. When in doubt, keep 'em around. They are your little crescent-shaped friends who will protect you from making mistakes. They help with negative numbers and exponents and all sorts of other things in pre-algebra and algebra. Numbers and variables running around "naked" can cause all sorts of problems.

Takeaway Tips

 When you're confused about what an exponential expression *means*, pay attention to parentheses and negative signs, and above all: When in doubt, multiply it out!

When a negative number is being raised to an exponent, if the exponent is odd, you will be left with a negative result. If the exponent is even, you can forget about the negative sign completely; it'll cancel! But above all: Look to see what's touching what, multiply things out as often as you'd like, and you'll be in great shape!

 When you put an exponent on an entire fraction, it's the same as putting the exponent on the top *and* bottom of the fraction.

TESTIMONIAL

JoAnna Kai Cobb (Bedford, Indiana)

<u>Before</u>: The target of gossip and dirty looks

<u>Today</u>: Singer, teacher, and web designer!

In elementary school, everyone wanted to excel at school. But in junior high, things changed. While a few of my friends still cared about school, we often felt like a secret society of sorts—a group of kids who studied in secret to protect our social identities!

But I loved math, and although I tried to ask my questions quietly at the end of class, the teacher would always end up saying, loudly, "JoAnna has an excellent question. I'd like everybody to write this down." Amid the moans and groans, I knew that I was again going to be the day's target of gossip and dirty looks.

> "I didn't like not being who I was simply to please others."

Finally, in eighth grade, I decided, *screw 'em*! I discovered that I didn't like not being who I was simply to please others. I started asking all the questions I wanted and ignoring the dirty looks. And you know what? Since I was asking the same questions that *they* were dying to ask (but were too afraid to), by the end of the year, my classmates would actually thank me for raising my hand.

Today, I'm a singer, a web designer, and a sixth-grade math teacher, and I love it all. Many people don't realize that web design is grounded in mathematics. I even use exponents in hexadecimal numbers to create my colors! *(Says Danica: For more on this, search "hexadecimal" at mathisfun.com.)* After my band performed at a local benefit concert recently, a former student introduced her father to me and said, "Dad, she's the reason I like math now!" Moments like these make me so happy, and I'm grateful for the wonderful variety in my life—largely thanks to math!

Chapter 16

Excuse Me, Have We Met Before?

Intro to Variables with Exponents

$\mathcal{I}$n the last chapter, we learned the basics of exponents and the perks of running your own company as a high-powered executive with a cute assistant named Brett. If you were impressed by the spur-of-the-moment champagne and caviar shindig, well, you ain't seen nothin' yet. Because now, we're graduating from numbers to *variables*.

Since exponents work the same no matter *what* the base number is—whether it's an integer, a fraction, a decimal, etc.—they also should work the same if we don't *know the value* of the base number, right?

Sound like a variable to you? Me, too.

Take a look at this:

$$y \cdot y \cdot y \cdot y \cdot y$$

Because there are 5 of the same thing being multiplied together, we can rewrite this as y^5. It doesn't matter that we don't know the value of y; we still need to *multiply it times itself five times*. Oh, that's so much neater looking, isn't it?

$$\underbrace{y \cdot y \cdot y \cdot y \cdot y}_{\text{5 of them}} = y^5$$

As you can see, when it comes to exponents, variables act the same way numbers do: The high-powered executive is still making short phone calls and making big things happen; the only difference with variables is that she's giving instructions to someone whose value she doesn't know. Nope, she doesn't know the person's name. Face not ringing a bell, either. Has she even met this person, working in her own office?

When it comes down to it, it doesn't really matter if she doesn't remember all her assistants' names, as long as she gives good instructions. Well, it would be more polite if she knew their names, but you get my point.* *Bottom Line: If Ms. Exponent is touching you, whoever you are, her instructions are to be carried out, regardless.*

Exponents and Parentheses

In the previous chapter, I mentioned how when you see an exponent touching the outside of a set of parentheses, then <u>whatever</u> is inside the parentheses will multiply times itself as many times as Ms. Exponent tells it to: "I don't care *who* you are. Multiply times yourself! Many times. Now!"

$$\left(-\tfrac{1}{2}\right)^6 = \left(-\tfrac{1}{2}\right)\left(-\tfrac{1}{2}\right)\left(-\tfrac{1}{2}\right)\left(-\tfrac{1}{2}\right)\left(-\tfrac{1}{2}\right)\left(-\tfrac{1}{2}\right) = \tfrac{1}{64}$$

$$(3x)^5 = (3x)(3x)(3x)(3x)(3x) = 243x^5$$

$$(\otimes)^3 = (\otimes)(\otimes)(\otimes)$$

$$(y+3)^4 = (y+3)(y+3)(y+3)(y+3)$$

Don't worry about how to multiply out that last one. Trust me, it wouldn't be pretty. I just wanted you to see it, so you can see what I mean when I say that "whatever" is inside really does multiply times its "whatever" self—no matter how complicated it looks!†

However, when the *only* operations inside parentheses are multiplication and division, this stuff doesn't have to get complicated.

. . . .

* Later, in advanced algebra, you'll see variables *as* exponents, like 2^x. That's when you don't know who your *boss* is. But we're not going to worry about that in this book.

† Just so you know, actually simplifying something like this (which requires a complicated use of the distributive property) is something you won't do until algebra.

Remember from p. 131 how we can only distribute multiplication over addition and subtraction? Well, Ms. Exponent has her own, stepped-up, "executive" version of this rule: You can *distribute exponents* inside a set of parentheses when the *only* operations inside the parentheses are multiplication and division*—NOT addition or subtraction.

In other words: For bases *a* and *b* and exponent *m*, we can distribute the exponent over multiplication:

$$(a \cdot b)^m = a^m \cdot b^m$$

. . . and also over division. (You might remember that we first saw this on p. 250):

$$\left(\frac{a}{b}\right)^m = \frac{a^m}{b^m}$$

Even when there's more than one base involved, the shortcut still works. For instance: $(a \cdot b \cdot c)^m = a^m \cdot b^m \cdot c^m$. In fact, no matter *how many* numbers or variables are inside the parentheses, as long as they're all stuck together with multiplication and division, everybody gets an exponent!

What this means for something like $(4xy)^3$ is that, rather than multiplying the whole thing out, we can distribute the 3 to everyone inside: $(4xy)^3 = 4^3 x^3 y^3 = 64x^3 y^3$.

It's always a good idea to know where these shortcuts come from, so let's figure this one out by expanding out $(4xy)^3 = (4xy)(4xy)(4xy)$.

This is just a bunch of multiplication, so we can write it in any order we want, right?

$$(4xy)(4xy)(4xy)$$
$$\rightarrow 4 \cdot x \cdot y \cdot 4 \cdot x \cdot y \cdot 4 \cdot x \cdot y$$
$$\rightarrow 4 \cdot 4 \cdot 4 \cdot x \cdot x \cdot x \cdot y \cdot y \cdot y$$
$$\rightarrow 4^3 \cdot x^3 \cdot y^3$$
$$\rightarrow 64x^3 y^3$$

And voilà!

. . . .

* Division in these kinds of expressions is almost always seen in fraction notation.

To see how this works for division (fractions), check out p. 250.

It sure is easier to follow the shortcuts, but it's nice to know why they work. Trust me, it can get dangerous when you start recklessly pulling exponents inside parentheses without understanding why!

"*I* think that it is good for people to show their abilities, and that's what smart girls do. They are proof that we as a society have moved past the era of silencing female opinion and intelligence. The more girls show their knowledge, the more people who have been hanging on to that horrible era will see how far we've come."
Johnathan, 14

Watch Out!

You can *only* distribute exponents over multiplication and division, not addition or subtraction: $(y + x)^2$ does NOT equal $y^2 + x^2$. So that you always avoid this mistake, I'm going to do a quick demonstration:

Let's evaluate $(2 + 3)^2$. Now, you *know* the answer is 25, right? After all, it's just 5 in there, and even if you wanted to "multiply it out," you'd still get 25:

$$(2 + 3)^2 = (2 + 3)(2 + 3) = (5)(5) = 25$$

But what if you had tried to distribute the exponent over the addition? You'd get:

$$(2 + 3)^2 \neq 2^2 + 3^2 = 4 + 9 = 13$$

Oops! And you can't do that with variables, either. That's why $(y + x)^2 \neq y^2 + x^2$. This is a mistake made again and again by algebra students. So, read this again! (I'm arching my eyebrows and staring ominously at you again . . .)

Doing the Math

Distribute these exponents inside the parentheses, *if possible*. If you're not allowed to distribute the exponent, then write "can't distribute." I'll do the first one for you.

1. $\left(\frac{2}{3}xy\right)^4$

<u>Working out the solution</u>: Because there is only multiplication and division inside the parentheses, it's safe to distribute the exponent to everyone inside!

So, $\left(\frac{2}{3}xy\right)^4 = \frac{2^4}{3^4}x^4y^4 = \frac{16}{81}x^4y^4$.

<u>Answer:</u> $\frac{16}{81}x^4y^4$

2. $\left(\frac{3}{4}x\right)^4$

3. $\left(\frac{3}{4}x + 5\right)^4$

4. $-\left(\frac{3}{4}xy\right)^4$ *(Hint: Notice that the exponent only affects stuff inside the parentheses.)*

5. $\left(-\frac{3}{4}x\right)^4$ *(Hint: Either write this out in expanded form so you can see what happens to the negative signs, or first "pull" out the -1 as a factor of $-\frac{3}{4}$ and rewrite it like this: $[(-1)\left(\frac{3}{4}x\right)]^4$. It's all multiplication/division, so <u>everybody</u> gets an exponent.)*

(Answers on p. 322)

See p. 317 in the Appendix for a handy chart of commonly used powers. It could be very helpful for homework.

And now it's time to say good-bye to Ms. Exponent. But she'll never be far . . . she'll always be smiling down at you from her corner office. And you'll feel the power you've gained from her as you move on to algebra and beyond!

Takeaway Tips

 Variables with exponents work the same way as numbers with exponents. Remember, a variable is just a number whose value we don't know yet.

When an exponent is outside of a set of parentheses, if the *only* operations inside are multiplication and division, you can distribute the exponent to everything inside!

TESTIMONIAL

Martha Tellez (Washington, D.C.)

<u>Before</u>: Panicked test taker
<u>Now</u>: Financial analyst at the Pentagon!

Throughout junior high and high school, I wasn't exactly a star student in math. English is not my first language, but I can't blame it on that, because by the end of elementary school, I seemed to be fine. For me, the problem was always the test taking. First there was the time issue. Knowing that I

> "Test after test, I failed..."

was up against the clock, I would panic and make careless errors. Sometimes I wouldn't even finish my exams. Then there were the nerves. I'd often get so stressed out before a test that I would lose the ability to think clearly. All of this amounted to less-than-stellar grades and a lack of confidence.

Then came calculus my senior year of high school. The class was notoriously difficult. After the first exam, which most students failed, almost half of the class transferred to an easier, slower-paced course. Not wanting to give up, I decided to stay in the class, even though I knew that it might hurt my GPA.

I had made it this far, and I wasn't about to give up now!

Test after test, I failed (or, at best, scored a C- or D+)—and not because I didn't understand the concepts, but because of my fear of tests! Over the course of the class, though, little by little—and through some serious hard work—I learned to take control of my fears and to replace panic with calm. Although I didn't exactly "ace" any of my exams, I did see improvements. I gave the class my all—and in return, I passed! I was so happy and relieved at the end of the semester. I did it!

Today, I'm a financial analyst for the Department of Defense, outside Washington, D.C. In a nutshell, I help to keep the U.S. Navy on budget. My office is at the Pentagon building, which is an amazing place to work because of its history and importance. I also love the variety of things I get to do through my job. I rotate through different divisions, and last year I spent four months on Navy ships in San Diego, which was awesome!

Math is an essential tool in the planning/ budgeting we do: My colleagues and I often run calculations to figure out how much money it will cost the government to build new ships for the navy, for example, as well as how much money in operation and maintenance these ships will require during their lifetime. In creating these estimates, we take into account the price of raw materials (wood, steel, etc.), modernization of the ships over time (most ships need to be updated every few years), the cost of the navy personnel that will occupy the ships, and so on. We must continually assess whether the projects we're working on will fit within our budget, and strategize about how to fund these projects. My work is exciting and challenging, and because my job is geared toward helping to keep the United States safe, I feel like I'm making a difference in the world!

I sincerely believe that discipline has gotten me where I am today. If you work hard and don't give up, you can accomplish anything you set your mind to.

You Said: Well . . . THAT Didn't Work!

Tried a new idea that flopped completely? Learned a lesson the hard way? See if any of these sound familiar!

"One time, I copied off a friend's homework, thinking it would save me tons of time. But then I was clueless during the test . . . I will never do that again!" **Levi, 12**

"Last year I had three essays due within a two-week period, and I thought that if I worked on all three of them simultaneously, I wouldn't get bored. Instead, two of the three ended up being really confusing. My English teacher could even tell that I was unfocused when I was writing the one for her class! I learned that it is way better to be thorough and approach assignments one at a time than to try and plow through them all at once." **Iris, 16**

"As a freshman, all I wanted was to fit in. Before I knew it, I was inviting people over after school, not doing my homework, sleeping during class to make up for the late hours I spent on the telephone, and skipping out on my family to hang out with people I barely knew. My grades were slipping fast, and I wasn't sure I'd ever get caught up! After a while, though, I realized that I didn't want to be the girl who made 'friends' in high school, only to lose them after graduation (if I even made it that far), and then be stuck with my poor excuse of an education. Well, no sooner did I make up my mind to change my ways than my grades started improving. It was hard, of course, but it made all the difference. I'm so much happier now!"
Stephanie, 17

"I admit that I am sometimes shy about asking questions during class. I just wait and hope that my specific questions will be asked by another classmate, and I never get around to asking. Then, on the test, there will be a question just like the one I didn't ask—and I'm stuck with no clue how to answer! My shyness needs to be resolved soon because I plan to become a teacher. I know there will be kids like me, and I want to help them get over this little fear." **Amy, 16**

"Last year was not a good year for me in math. I wasn't understanding anything, so I just gave up completely. By doing that, I missed out on a year of math and had to work double hard to catch up!" **Jenna, 17**

"Has anyone ever sweet-talked you into completing one of their assignments? I was in my last class of the day, history, trying to finish my Spanish homework so I wouldn't have to carry my Spanish book home. One of my friends asked me if I could finish his Spanish for him, too. I said yes out of the kindness of my heart, but the bell rang before I finished. I quickly stuffed all of my papers into my bag and went home. The next morning, when I went to get my homework out of my Spanish notebook, it wasn't there—I had accidentally left it in my history notebook! Of course, I decided to give the completed assignment to my friend to turn in. So my friend got an A, and I took a zero. I learned from my mistake, and it was the last time I ever finished anyone's homework besides my own." **Brittany, 15**

"My locker used to be a mess. I jammed papers into my books, and I couldn't find anything. Then one day I lost a really important assignment! I couldn't find it, so I wasn't able to turn it in. I received an F on that assignment, and I learned a major lesson in organization. Habits are hard to break, but I now try to clean my locker regularly so I don't lose stuff!" **Desiree, 13**

"We were doing this hard worksheet in social studies, and my friend's really smart, so I asked if I could copy hers. She said I could, so I thought that was pretty great, but what I didn't know was there was a pop test coming! When I sat down in class the next day and saw 'TEST TODAY' on the board, I started to worry. Looking over it, I saw that it covered all the information from the day before! I started guessing all the answers, but I ended up getting what I deserved on that test: a big, red F. I learned my lesson—I'll never copy off someone else's work again." **Brittany, 14**

Do You Sudoku?

I was first introduced to Sudoku on the set of *How I Met Your Mother*. (See pp. 190–1 for a testimonial from Executive Producer/ Director Pam Fryman!) During the taping of my first episode of the series, Alyson Hannigan (character: Lily) taught me how to play. I quickly discovered that the whole set was obsessed with it—actors, crew members, you name it! Whenever they're not taping, out come the Sudoku puzzles.

Here are the basic rules: The goal is to fill in each square with a number 1 through 9, so that every column has exactly one of each digit (no repeats), every row has exactly one of each digit, and every bolded 3 × 3 box has exactly one of each digit. The challenge is that every row, column, and 3 × 3 box must have *one and only one* of each digit.

1	3		5					
		8	7		9	6		
	7			6		3		2
				5	3		2	4
6		4	9		7	8		5
2	5		1	4				
7		9		1			4	
		2	8		4	7		
3					6		9	

Wanna try? Let's look at the middle horizontal row—the one that starts with a 6 and ends with a 5. It only has three open spots, and if you look at the digits already in that row, you'll see that the only digits "needed" in the row are 1, 2, and 3. Looking at the first empty spot (the one shaded gray), let's see if we can figure out what goes there. Checking its column, there's already a 3 (up top). We can't have another 3 in the column, so that's out. Also, within its 3 × 3 box, we can see that there's already a 2, and since we can't repeat the same digit within a bolded 3 × 3 box, the spot can only be filled in with a 1! Go ahead, write it in. Tip: Never write in a number unless you know it *has* to be true. Also, be sure you've finished your homework before you do these; they can be addicting! For more puzzles and strategies, do an Internet search for "easy Sudoku," and for the solution to this one, see kissmymath.com. Good luck!

Chapter 17

Secret Sausages
Intro to Functions

Have you seen something like this in your textbook yet?

$$f(x) = x + 1$$

If so, did you think to yourself, "Now, what the heck is that?"

Well, whether you have or not, I have good news: Soon, not only are you going to understand just what the heck "$f(x)$" is, it's going to be easy to handle and will make *graphing lines* a breeze. (More on that in Chapter 18).

And remember all that "substituting for x" we did back in Chapter 6? Well, it's about to come in handy!

THE SAUSAGE FACTORY

Understanding *Functions*—What Are They, Anyway?

Functions are often written like this: $f(x) =$ "something involving x."
Examples would be:

$$f(x) = x + 2$$
$$f(x) = \frac{(3x - 1)}{5}$$

But what does any of that mean? Well, the f in $f(x)$ stands for the word *function*, but I like to think of it as *factory*. And what does this function or factory do? It takes IN the value of *x* and puts OUT some other value, which we call: *f(x)*.

So, we put *x* into this factory, and we get some other value at the end. Different values of *x* that go INTO the sausage factory will give us different OUT values of *f(x)*. But no matter what ingredient you put INTO the factory, the factory will still do the *same things* to it.

This makes sense: After all, we can imagine that in a sausage factory, machines might grind up the meat, sprinkle it with pepper, add some onions, and then chop it up. But the sausages that come out are going to taste very different, depending on which meat you put in, right? If you put in chicken, the sausage will taste different than if you put in beef, even though the factory did the exact same things to each meat.

If no one were looking, you could put in a candy bar and see what you'd get. Let's see, it would be ground up, sprinkled with pepper, have onions added to it, and then be chopped up. Yuck! It wouldn't taste very good, but you could do it!

Let's put two different ingredients into the function $f(x) = 3x + 2$ and see what happens. We know that no matter which value of *x* goes into the factory, first it will be multiplied by 3, and then it will get 2 added to it.

First, let's see what would happen if we tried $x = 1$.

$$f(x) = 3x + 2$$
$$\rightarrow f(1) = 3(1) + 2$$
$$\rightarrow f(1) = 3 + 2$$
$$\rightarrow f(1) = 5$$

So we've discovered that for the IN value of $x = 1$, the sausage factory gives us the OUT value of 5. In other words, $f(1) = 5$.

Now let's see what happens if $x = 20$.

$$f(x) = 3x + 2$$
$$\rightarrow f(20) = 3(20) + 2$$
$$\rightarrow f(20) = 60 + 2$$
$$\rightarrow f(20) = 62$$

This time, with the ingredient $x = 20$, we get $f(20) = 62$.

Ring Ring **What's It Called?**

Function

A **function** is a type of equation that represents instructions for what to "do" to x. You can think of it like a factory that receives an ingredient, x, and delivers a sausage, $f(x)$. Functions are often written with $f(x)$, or sometimes just with y. Here are some examples of functions:

$$f(x) = 3x - 1 \qquad f(x) = x^2 \qquad y = 2x + 1$$

Note that for each ingredient, a function can deliver only one kind of sausage.*

- - - - - - - - - - -

* This is a part of the definition of *function* that's nitpicky but important later on. See, there are equations called "relations" where you *can* have more than one sausage for each ingredient—more than one output for each input. (Very un-factory-like, don't you think?) We'll also talk about it a bit in the WHAT'S THE DEAL? on p. 301, but you probably won't see much of these relations until later in algebra.

Using Tables

Yes, it's most polite to eat your sausage at the table . . . but that's a different kind of table. For the function $f(x) = 2x + 1$, let's use the ingredients $-2, -1, 0, 1, 2$ and see what kinds of "sausages" we get for each input. We'll keep track of them in the table below!

$$f(x) = 2x + 1$$

Ingredient: x	→	Sausage: f(x)
−2	→	−3
−1	→	−1
0	→	1
1	→	3
2	→	5

Showing the work

$f(-2) = 2(-2) + 1 = -4 + 1 = \boxed{-3}$

$f(-1) = 2(-1) + 1 = -2 + 1 = \boxed{-1}$

$f(0) = 2(0) + 1 = 0 + 1 = \boxed{1}$

$f(1) = 2(1) + 1 = 2 + 1 = \boxed{3}$

$f(2) = 2(2) + 1 = 4 + 1 = \boxed{5}$

So all of the ingredients (values of x) we plugged in are on the left side, and the answers we get after plugging them in are on the right. No matter what number goes into this factory, it will be doubled and then added to 1. When we put a 0 ingredient into the factory, we get a 1 sausage. When we put a 2 ingredient into the factory, we get out a 5 sausage. See?

You don't have to use those arrows when you make this kind of table, but I like them because they show us that we started out with x, and then, after the factory "did its thing" to x → we ended up with the $f(x)$ value. Making tables will be very helpful once we get into graphing.

Here's another kind of table . . .

Whispering the Secret Sausages: Another Kind of Table

Whenever I see something written in parentheses, I always think of someone whispering. (See how you can make the "reading voice" in your head whisper, every time you see something in parentheses?)

Imagine that you found a hidden factory that makes the most delicious and unexpected sausages from its ingredients, and you want to tell a friend about its secret recipes. You'd crouch down next to her and whisper in her ear: "(Shhh! Don't tell anyone, but when I put oranges into this factory, the sausage ends up tasting like chocolate!)" or "(When I put in a 2, I get out a 163!)."

If you didn't have much time, you might whisper a much shorter phrase like, "(oranges, chocolate)" or "(2, 163)," and your friend would know what you mean, because the two of you had already decided on the *order*: You'd always whisper the ingredient first, and *then* the resulting sausage. Pretty efficient, huh?

Let's write down our function table again for $f(x) = 2x + 1$, but instead of using arrows, this time let's "whisper" the ingredient/sausage pairs in little parentheses, as if you were sharing a secret. Here's the table from p. 268, in "whispering" form. Shhh . . .

(ingredient, sausage)
(−2, −3)
(−1, −1)
(0, 1)
(1, 3)
(2, 5)

Step By Step

Evaluating functions and making tables of the results:

Step 1. Take each ingredient (each specific value for *x*) and substitute it for *x* in the function. Then simplify the expression to find out $f(x)$'s value.

Step 2. Do this for all specified values of *x*, and keep track of the results you get.

Step 3. Make a table that organizes these results. Put all the *ingredients* (specific values of *x*) on the left, and their corresponding *sausages* [values of *f(x)*] directly to their right. You can either use arrows in the table or use the super-efficient whispering parentheses form: "(ingredient, sausage)."

For Step 3, your book or teacher might indicate what kind of table to draw, but the "whispering parentheses" table will be more helpful as we get into graphing. You'll see why in the next chapter!*

QUICK NOTE   When you plug values for x into your factories and then simplify, it's just like when we substituted different values for variables on pp. 84–5. So, as always, be sure to use the PEMDAS Order of Operations rules while you're simplifying.

Step By Step In Action

Let's evaluate the function $f(x) = 8x - 1$ *at these values of x:* -2, 0, $\frac{1}{2}$, *and* 5. *We'll make both kinds of tables to summarize our results. Let's begin!*

Steps 1 and 2. Plug in each value of *x*, and see what value of *f(x)* we get. For $x = -2$, we'll plug in **–2** everywhere we see *x* and get $f(-2) = 8(-2) - 1 \rightarrow f(-2) = -16 - 1 \rightarrow \textbf{f(-2) = -17}$. Okay, we've found our first ingredient/sausage pair. We put in –2, and we got out –17. Next, for $x = \textbf{0}$, we get $f(0) = 8(0) - 1 \rightarrow f(0) = 0 - 1 \rightarrow \textbf{f(0) = -1}$. Got our next combo: We put in 0, we got out –1. Alright, next ingredient: let's say $x = \frac{1}{2}$. Then $f\left(\frac{1}{2}\right) = 8\left(\frac{1}{2}\right) - 1 \rightarrow f\left(\frac{1}{2}\right) = 4 - 1 \rightarrow \textbf{f}\left(\frac{1}{2}\right) \textbf{= 3}$. Okay: We put in $\frac{1}{2}$, we got out 3. What happens if $x = 5$? Let's see what kind of sausage our factory will produce from it:

$$f(\textbf{5}) = 8(5) - 1 \rightarrow f(5) = 40 - 1 \rightarrow \textbf{f(5) = 39}$$

Step 3. Now we'll summarize our results by putting them in tables. First, we'll draw the table that uses an *arrow* from each ingredient to the

. . . .

* We'll define *whispering parentheses* as "ordered pairs" in Chapter 18, and that's how your teacher will know them. But *we* know about the secret sausages, don't we?

sausage that goes with it. Then, we'll make an even more efficient table with the (ingredient, sausage) pairs using *parentheses*:

ingredient: x	→	sausage: f(x)
−2	→	−17
0	→	−1
$\frac{1}{2}$	→	3
5	→	39

(x, f(x))
(−2, −17)
(0, −1)
($\frac{1}{2}$, 3)
(5, 39)

QUICK (REMINDER) NOTE   Don't worry about what these tables "mean." Sure, tables with arrows and whispering parentheses will be helpful later on when we want to graph functions; but for now, just think of them as ways of writing down our answers in an organized way.

 Doing the Math

For each function, evaluate f(−6), f(−3), f(0), f(3), and f(9). In other words, what does f(x) equal, when x equals −6, −3, 0, 3, and 9? Then draw both kinds of tables to organize your answer: with arrows and with "whispering" parentheses. I'll do the first one for you.

1. $f(x) = 3x + \frac{x}{3} + 1$

Working out the solution: Let's evaluate f(x) at each value of x we've been given: For x = −6, we get
$f(-6) = 3(-6) + \frac{-6}{3} + 1$ →
$f(-6) = -18 + (-2) + 1$ → $f(-6) = -20 + 1$
→ $f(-6) = -19$. $f(-6) = -19$

For x = –3, we get $f(-3) = 3(-3) + \frac{-3}{3} + 1 \to$
$f(-3) = -9 + (-1) + 1 \to f(-3) = -9.$ $\mathbf{f(-3) = -9}$

For x = 0, we get $f(0) = 3(0) + \frac{0}{3} + 1 \to$
$f(0) = 0 + 0 + 1 \to f(0) = 1.$ $\mathbf{f(0) = 1}$

For x = 3, we get $f(3) = 3(3) + \frac{3}{3} + 1 \to$
$f(3) = 9 + 1 + 1 \to f(3) = 11.$ $\mathbf{f(3) = 11}$

For x = 9, we get: $f(9) = 3(9) + \frac{9}{3} + 1 \to$
$f(9) = 27 + 3 + 1 \to f(9) = 31.$ $\mathbf{f(9) = 31}$

Now we can arrange our answers in tables!

Answer:

$x \to f(x)$	$(x, f(x))$
$-6 \to -19$	$(-6, -19)$
$-3 \to -9$	$(-3, -9)$
$0 \to 1$	$(0, 1)$
$3 \to 11$	$(3, 11)$
$9 \to 31$	$(9, 31)$

2. $f(x) = 2x - 2$

3. $f(x) = \frac{2x}{3} - 3$

4. $f(x) = x^2 - x$

(Answers on p. 323)

"*I* didn't used to like math; it seemed like a big jumbled mess. Now math is my favorite subject. I think that this year might be tough, but I'll hang in there!" **Tiffany, 14**

Moving Toward Graphing Lines: y Instead of $f(x)$

There are two ways of writing functions: You can use $f(x)$ or just y. And, as you get into graphing lines, you'll start seeing functions written using a y *instead* of $f(x)$. Everything else is exactly the same!

For example, $f(x) = x + 3$ and $y = x + 3$ are the same function. Let's practice calling the sausage y, so you can get used to it. When you think about it, the letter y looks more like a sausage than $f(x)$ does, anyway. Well, maybe only a little.

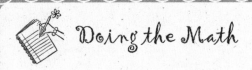 Doing the Math

For each function, evaluate y when $x = -4, -1, 0$, and 3. Write out the "whispering" parentheses along the way to keep track of the (ingredient, sausage) pairs. Then, make both kinds of tables to summarize your results. I'll do the first one for you.

1. $y = 8 - x$

<u>Working out the solution:</u>

For x = **-4**, we get
y = 8 - (**-4**) → y = 8 + 4 → y = 12.
Alright, for our first whispering parentheses, we got (**-4, 12**).

For x = **-1**, we get
y = 8 - (**-1**) → y = 8 + 1 → y = 9.
So our next pair is (**-1, 9**).

For x = **0**, we get y = 8 - (**0**) → y = 8.
Next whispering pair: (**0, 8**).

For x = **3**, we get y = 8 - (**3**) → y = 5.
Our final ingredient/sausage pair is (**3, 5**).

Arranging them into tables, we get:

Ingredient → Sausage

x → y	(x, y)
-4 → 12	(-4, 12)
-1 → 9	(-1, 9)
0 → 8	(0, 8)
3 → 5	(3, 5)

2. $y = x + 1$

3. $y = 6 - x$

4. $y = 4x - 5$

(Answers on p. 323)

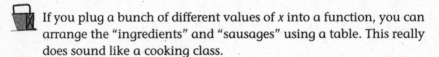

Takeaway Tips

A *function* is a type of equation that represents instructions for what to "do" to x. You can think of it like a factory that receives an ingredient, x, and delivers a sausage, $f(x)$. Functions are often written with $f(x)$, or sometimes just with y.

If you plug a bunch of different values of x into a function, you can arrange the "ingredients" and "sausages" using a table. This really does sound like a cooking class.

Using "whispering" parentheses to record the (ingredient, sausage) pairs is a great way to organize your results when you plug values of x into a function. In Chapter 18, we'll learn that these (x, y) pairs are called *ordered pairs*. Then we'll learn how to *graph* these pairs!

Danica's Diary

STOP, DROP, AND MAKE A LIST!

Got too much to do? We've all been there. I used to get really stressed whenever I had a lot of homework. What should I work on first? The big English paper, my science report, or studying for that huge math test? All too often, it seemed like I had *too many things to do.* I had no idea how to do them all, and most of the time, I'd feel paralyzed by this, and wouldn't even know where to start. Oddly, sometimes having too much to do would make me procrastinate *more*, because I was afraid of facing it at all.

MAKE A LIST!

Here's what I learned over the years: When you come home from school and you have too many things to do, stop, take a breath, and...make a list.

Make the list as *long* as you want. It can include all your assignments and anything else you have to do, even things like taking out the trash and eating dinner. Seriously, I still do this, and sometimes I'll include things like "take a shower" or "eat," along with "write five pages today for KMM."* It's just a list, and you're the only one who sees it, after all! You can also break down some items. For instance, "MATH" can become two items on the list: "Do problem set" and "Start Pretend Cheat Sheet[†] for next Friday's test." Then, as objectively as you can and while breathing calmly, circle the things that *need* to get done tonight—the biggest priorities.

Okay, now at this point you might say to me, "But even just the *circled* things will take, like, 20 hours, and I only have one night!" That's okay. Maybe you're right, and maybe you're wrong. I'm now going to tell you the one, single secret to doing

. . . .

* *Kiss My Math*

† See p. 306 for the Pretend Cheat Sheet! It makes studying for tests *sooo* much easier.

work when there's *too much*, when you're feeling the
pressure of the clock, and when you don't know where
to begin. It comes down to one word:
  Start.
  Yes, start. Look at your high-priority, circled
list, and just start. Somewhere. Anywhere. Which
thing on the list should you start with, you ask?
Any thing—so long as you start! Don't worry, don't
think, just start. The more you have to do, the more
important this is. Isn't that great news?
  Finally, here's the fun part about having a list:
You get to *check things off* when you do them. I like
to start with important but shorter items, so that I
can start checking things off sooner. It feels good
because you can see progress happening.
  You may still not know how you're going to finish
it all, but you know that you're *closer*. Resist the
urge to worry. Just keep moving forward in a sort of
ignorant bliss. Don't look down when you're walking
the tightrope; just keep putting one foot in front
of the other.
  If one task is not going well and you know you're
being unproductive, it's okay to put it down and
move on to another item on your list. I have a
little symbol that I put next to things that I
started but didn't finish: It's like a half-check
mark, just a little slanted line. You can always
come back to those items when you're so inspired.
  As I mentioned, if I have a lot to do, I'll make
a list for the day. But sometimes I'll make a list
for the weekend or even for the whole week if things
are really busy. And I sure do like checking things
off. If I'm in a grumpy mood, I might even put a big
X through the item that I just finished. Either way,
the checking-off process is very satisfying. I
really get a feeling of accomplishment from it.
  Learning how to balance workloads and manage your
time (and your stress level) is one of the most
important things you can learn from school, because
no matter what career you choose, these skills will
help you throughout your *entire* life. So go ahead...
make a list!

Two Lovely Ladies

$\mathscr{P}$erhaps more than we realize, there are fabulous women all around us who use math in their jobs. As I was thinking about my life, I realized that my sister and my best friend are two of these women, so I asked them to share with you how math has benefited them in their jobs!

Crystal McKellar,
Lawyer in New York City

Most people think that being a lawyer has nothing to do with math. In fact, I remember college friends saying they wanted to pursue law careers rather than careers in consulting or investment banking because they hated math. However, lawyers who are comfortable with math are at a huge advantage!

My law firm represents companies, and most of the time companies need our lawyers' help because someone is suing them over money or other assets. An asset is something that a company owns that is expected to bring in income. For example, a candy company might have as its largest asset a candy factory, and somebody might sue the candy company, saying "You're not being truthful about the value of that factory!" And we can say back, "Oh really? We'll just see about *that*."

One way for determining the value of any asset, including candy factories, is via a method called a *discounted cash flow method*. The formula looks like this:

$$\text{Discounted Cash Flow for 3 years} = \frac{CF_1}{(1 + r)^1} + \frac{CF_2}{(1 + r)^2} + \frac{CF_3}{(1 + r)^3}$$

where CF = the asset's cash flow for 1st, 2nd, or 3rd years, and r = discount rate.

Don't worry. None of this math should make any sense to you now, but with an understanding of exponents and basic algebra *now*, someday later you'll be able to understand this stuff without a problem!

Many lawyers do not take as much math as I did, which is fine. But I took economics, accounting, and finance in college and law school, which are courses that all involve math. Because

I have a particularly strong handle on the math that is at the core of the companies' legal disputes, I can discuss the issues intelligently with the senior partners, and that can help us determine our strategy for winning our case. It's a great feeling!

Kimberly Stern,
Director of Operations in
health care in Los Angeles

I work for a company that provides *lifesaving* in-home dialysis treatments to people with kidney failure. Math isn't just important in our day-to-day operations—it's critical!

I love my job. I get to travel all over the country every month, working with physicians, nurses, analysts, and strategists. It's exciting! I'm constantly running from meeting to meeting to brainstorm different ways to get the word out to physicians, dialysis patients, and their families about this liberating alternative to in-center dialysis and how it can improve patients' quality of life. And yes, I use math.

One of the math formulas my company depends on is for something called CAGR, Compounded Annual Growth Rate. By being able to see how the company has grown over a recent period of time, this formula allows us to budget and plan for things like staffing, salaries, and marketing materials. (Later, in algebra, you'll learn about having exponents that *are* fractions—and variables!)

$$\text{CAGR} = \left(\frac{Ending\ value}{Beginning\ value} \right)^{\frac{1}{x}} - 1$$

where x = the number of years we're analyzing.

The math I use is no higher than algebra II, but every day, I use the critical problem-solving skills that I first developed in math classes like algebra and calculus. That's just what math does. If you practice the more challenging problems today, you will be better prepared to analyze and solve real-life problems tomorrow. You never know: Someone's life may depend on it!

Chapter 18

Creative Uses for Bubblegum

Intro to Graphing Points and Lines

I'm not a big bubblegum fan. I don't like artificial sweeteners, and it's almost impossible to find gum without them! Besides, it makes my jaw tired. Anyway, there are other good uses for gum, so all is not lost. I'll get to that in a few pages.

First, we're going to learn how to graph points on a plane. For starters, you have to be on an airplane. (Just kidding.) Seriously, remember how we can graph numbers on a line like we did on p. 13? For example, if we wanted to graph the numbers -4, -1.5, and 3, it would look like this:

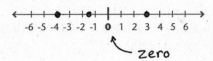

But that was *so* one-dimensional . . . literally. Now we can graph in *two* dimensions, on something called a *coordinate plane.** It's made up of two regular 'ol number lines, but one of them is sideways, and they cross at their zeros. See?

The place where they cross is called the *origin*, I guess because zero is where everything starts, or "originates." The *x*-axis is just like the number line: Numbers to the right of the origin are positive, and numbers to the left of the origin are negative.

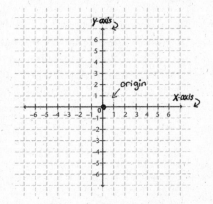

. . . .

* Sometimes it's called the Cartesian plane.

The *y-axis* is the line that goes up and down, and it's pretty easy, too: Numbers above the origin are positive, and numbers below are negative. With me so far?

Also, the two axes* divide the plane into four parts called "quadrants." Just in case you have to memorize where the quadrants are: Pretend like there's a little acrobat holding onto a rope that's tied to the origin (don't try this at home). She starts off in the positive quadrant, because . . . it's so happy there. Then, she runs along the *x-axis* a few steps, toward the *y-axis*, *jumps* up, and swings around the whole thing, hanging onto the rope and yelling "Wheeee!"

This would cause her to swing through them in order—Quadrant I, II, III, and IV—and end up back in Quadrant I again, ready for more swinging!

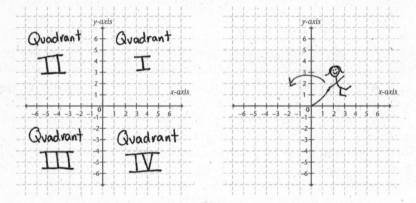

Tell me: If I made you label the four quadrants right now on a blank coordinate plane, could you do it? Yep, that's what I thought. You're a total stud.

So, if we graph normal numbers as points on a one-dimensional *line*, how do we graph points on a *plane*? We have to use *pairs* of numbers!

On p. 269, we whispered secret sausage recipes like this: (ingredient, sausage).† As it turns out, pairs of numbers written like that, for example (2, 3) and (−4, 1), are called *ordered pairs*. And *each ordered pair can be plotted as a single point on our two-dimensional plane!*

. . . .

* *Axes* is the plural of *axis*. I think it's a Latin thing. Also, *axes* rhymes with *taxis*, another plural word. *Axis*, however, rhymes with something you might say to your sister before a vacation if her suitcases were still empty: "Pack, sis!"
† If you haven't already, I recommend reading about functions in Chapter 17. It makes graphing *so* much easier, and it won't take you very long—promise!

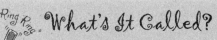

"Ring Ring": What's It Called?

Coordinate Plane

A **coordinate plane** is a two-dimensional surface, divided into four quadrants, where we can plot points (and eventually graph lines and other functions, too).

Ordered Pair

An **ordered pair** is two numbers inside parentheses, separated by a comma, which together represent a *single point* on the coordinate plane. Examples of *ordered pairs*: (1, 2), $\left(-3, \frac{1}{2}\right)$, (0, 0). Ordered pairs take the form (x, y), where the first number is the x value, also called the **x-coordinate**, and the second number is the y value, also called the **y-coordinate**. The x-coordinates in the above ordered pairs are 1, -3, and 0. The y-coordinates are 2, $\frac{1}{2}$, and 0. It's easy to remember that the x-coordinate comes first in ordered pairs, because x comes before y in the alphabet!

Plotting Points

On p. 269, we whispered pairs of values (oranges, chocolate) and (2, 163) in a specific *order*: first the "ingredient" and then the "sausage." When we plot points, we also follow a specific order: Starting with our finger on the origin, the *first* number tells us <u>how far along the x-axis we should go</u>, and the *second* number tells us <u>how far up or down the y-axis we should go</u>.

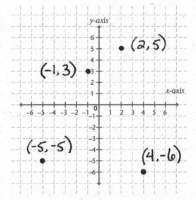

So, to plot the point (2, 5), since the x-coordinate is positive 2, we start with our finger on the origin and move it along the x-axis in the *positive* direction 2 units. Then, because the y-coordinate is positive 5, we move *up* 5 units and draw our point there!

To plot the point $(-1, 3)$, since the x-coordinate is negative, again starting at the origin, we first move to the *left* (negative direction) 1 unit; then, we move *up* (positive direction) 3 units. On the other hand, for a point like (4, -6), we first move to the *right* (positive) 4 units; then, because the y-value is negative, we move *down* 6 units. If both the x-value and the *y*-value are negative, like (−5, −5), we first move to the left and then down. Not so bad, right?

QUICK NOTE   The ordered pair (0, 0) represents the origin—the point where the x-axis and y-axis cross.

QUICK NOTE   If 0 is the first number (the x-value) in an ordered pair, the point will lie *on the y-axis*. For example, (0, 1) and (0, −8) lie *on the y-axis*. If 0 is the second number (the y-value) in an ordered pair like in (11, 0) and (−5, 0), then the point will lie *on the x-axis*. Try plotting them and see for yourself!

Doing the Math

Plot these ordered pairs as points on a coordinate plane. I recommend using graph paper, if you have it.* Also, label the quadrants I, II, III, and IV. I'll do the first one for you.

1. (−1, −4), (0, 2), (2, 6)

<u>Working out the solution</u>: To plot the point (−1, −4), first we'll move in the negative direction along the x-axis 1 unit and then down 4 units. Next, for (0, 2), we'll first move "0" units along the x-axis. What does that mean? It means we're still stuck at

* You can download and print graph paper from www.printfreegraphpaper.com or http://themathworksheetsite.com/coordinate_plane.html, which already has the *x*-axis and *y*-axis drawn for you, too!

the origin! Next, we look at the y-coordinate to see how much we should move up or down. It's a positive 2, so we go up 2 units. For the third point, (2, 6), it's all positive, so we just go to the right 2 units and then up 6 units. To label the quadrants, we just need to imagine the little acrobat (see p. 280) and we're golden!

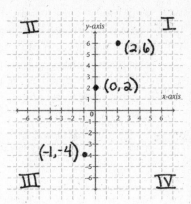

2.   (3, 4), (−5, −6), (−1, 0), (−4, 5)

3.   (2, −3), (6, −3), (4, 5) *(Connect the dots for a triangle!)*

4.   (2, 0), (0, 2), (−2, 0), (0, −2) *(Connect the dots for a diamond!)*

(Answers on pp. 323–4)

QUICK NOTE   Although we've been working mostly with integers (positive and negative whole numbers), the x-axis and y-axis each represent the entire set of real numbers,* so points like $(\pi, 1)$ and $\left(3.57, -\frac{1}{8}\right)$, also exist on the plane. It may be tough to plot them accurately, but they're there!

How to Use *Functions* to Create Ordered Pairs—and Lines!

We worked with functions (you know, sausage factories) in Chapter 17. Examples of functions are $f(x) = 3 - x$, and $y = x + 2$, remember?

. . . .

* See p. 314 in the Appendix for the definition of "the set of real numbers."

Well, sometimes we may want to see a "picture" of a function. After all, from a function, we can create a table of (x, y) pairs, as we did on p. 274. What would a collection of (x, y) pairs *look like* on a graph?

We've been plotting points, right? Well, instead of me just giving you a random list of ordered pairs to plot, I could instead ask you to *create* your own list of ordered pairs—from a function!

On p. 274, look at the table on the right. Do you see how those (ingredient, sausage) pairs are actually ordered pairs that you could plot on a plane? Let's do it and see what these functions *look* like.

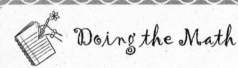 *Doing the Math*

a. Evaluate each *function* at the following *x*-values, and make two tables: one using arrows and the other using "whispering" parentheses, which of course are ordered pairs!
b. Then, graph each *ordered pair*. These functions all happen to be <u>linear</u>, so if you plot the points correctly, they'll connect to make a straight <u>line</u>! I'll do the first one for you.

1. $y = 2x + 1$, where $x = -3, 0$, and 2

<u>Working out the solution:</u> Part **a** is like what we did on pp. 273-4 in Chapter 17, so:

For x = −3, we get
$y = 2(-3) + 1 \rightarrow y = -6 + 1 \rightarrow y = -5$.
So our first pair is **(−3, −5)**.
For x = 0, we get
$y = 2(0) + 1 \rightarrow y = 0 + 1 \rightarrow y = 1$.
Our next ordered pair is **(0, 1)**.
For x = 2, we get
$y = 2(2) + 1 \rightarrow y = 4 + 1 \rightarrow y = 5$.
And our last pair is **(2, 5)**.

To finish part **a** and do part **b**, we'll arrange these three points in tables and then plot them.

<u>Answer:</u>

a.

Ingredient → Sausage

X → y	(x, y)
-3 → -5	(-3, -5)
0 → 1	(0, 1)
2 → 5	(2, 5)

b.

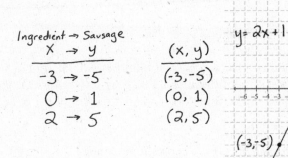

$y = 2x + 1$

2. $y = x + 3$, where $x = -3, 0,$ and 3

3. $y = 3x - 2$, where $x = -1, 0,$ and 1

4. $y = 1 - x$, where $x = -2, 0, 1,$ and 4

5. $y = x + 3$, where $x =$ three random numbers that YOU choose. *(Hint: No matter what values of x you pick—seriously, pick any numbers you want—if you plug them into the equation correctly, you'll end up drawing the same line that you drew in #2; however, pick numbers not too far from zero so the points will fit on your graph paper!)*

(Answers on pp. 324–5)

Graphing Lines

Say somebody gave you an equation like $y = 5x + 2$, didn't say anything else, and told you to graph the line. What's a girl to do?

Actually, you know *exactly* what to do! Just pick a few numbers—*any* numbers—and plug them into the equation wherever you see x. Find out what y value you get for each of them, and make a small table of your

(x, y) points. Then graph those points and draw a line through them. Done!

Step By Step

Graphing lines from equations: the "plug and plot" method:

Step 1. Pick three values* for x. Yep, *any* numbers will do (but $x = 0$ should always be one of them because it's so easy to work with)! Plug them into the equation of the function, and create a small list of (x, y) ordered pairs.

Step 2. Plot these ordered pairs on a coordinate plane.

Step 3. Using a ruler, draw a perfectly straight line through them. Done!

QUICK NOTE   If you don't have a ruler around, you can draw straight lines using anything sturdy and straight like the edge of an index card. Even a long nail file would do the trick!

 **And... Step By Step In Action Action!**

Graph the line given by the equation: $y = \frac{1}{3}x - 2$.

Step 1. Let's pick some values to plug in for x. Hmm, what looks easy? Well, 0 is always easy, so if $x = $ **0**, then $y = \frac{1}{3}$(**0**) $- 2 \rightarrow y = -2$. Okay,

• • • • • • • • • •

* Technically, a line is defined by two points, so you only *need* to graph two points and then draw a straight line through them. But I recommend graphing three points; in a way it's actually faster. If you've done the math correctly, you should be able to draw a *straight line* through all three points. It's the easiest way to check your work without having to redo anything.

so our first point will be **(0, −2)**. Next, let's use $x = $ **3**, because it'll make the fraction go away: $y = \frac{1}{3}(3) - 2 \rightarrow y = \frac{1}{3} \cdot \frac{3}{1} - 2 \rightarrow$ (canceling the 3's) $\rightarrow y = 1 - 2 \rightarrow y = -1$. So our new point will be **(3, −1)**. For our third value of x . . . well, let's see. Something else divisible by 3 would be nice, so the fraction will go away again. How about $x = 6$? Plugging $x = $ **6** into the equation, we get $y = \frac{1}{3}(6) - 2 \rightarrow y = \frac{1}{3} \cdot \frac{6}{1} - 2 \rightarrow$ (canceling factors of 3) $\rightarrow y = 2 - 2 \rightarrow y = 0$. This means we've discovered that another point on the line is **(6, 0)**. Our list of ordered pairs is $(0, -2)$, $(3, -1)$, and $(6, 0)$.

Steps 2 and 3. Time to plot and draw the line!

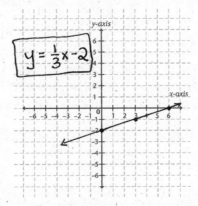

And that's one way to graph a line: Just plug in a few x values and collect (x, y) points to graph. And voilà! I'll bet you didn't realize you already knew how to do that, huh?

What's the Deal?

Graphing Linear Functions as Lines: What Does It All Mean?

Let's take a moment to get a little perspective and think about what we've been doing. Each function that we've been graphing has been a *linear* function. This just means that their graphs all look like straight *lines* when we connect their points.

When we look at a linear function like $y = x + 3$, what we're actually looking at is a <u>formula</u> for how to find every single (x, y) point on this line. In fact, no matter what value of x we stick in, the function will deliver the correct y value so that the (x, y) pair describes a point that falls *directly* on the line! This is because:

The equation of a line describes a particular <u>relationship</u> between x and y for all the (x, y) pairs, so *all the points* that the line goes through will share that same <u>relationship</u> between their x-coordinates and y-coordinates.

For example, when we graph the line $y = x + 3$, we'll end up looking at the line of <u>all the points on the plane whose y value is 3 *more* than its x value</u>: $(2, 5)$, $(3, 6)$, $(100, 103)$, and so on. Yep, they'll all be on the line!

Here's another way to think about lines and their equations. Remember in Chapter 12 how we solved equations like $5 - x = 3$, and we found that $x = 2$ was the one value of x that made the equation a *true statement*? Well, with an equation like $y = x + 3$, the line passes through <u>all the pairs of values (x, y) that make the equation a *true statement*</u>!

This is a lot of new information to digest, so feel free to reread this box a bunch of times. Don't expect to "get" it all; just simply *read* it. Slowly it will seep into your subconscious. Then, at some point in the not-so-distant future, it will all just click! When you understand the stuff in this box, you'll have a deeper comprehension of lines and their graphs than most people *ever* do . . . and you'll be a superstar by the time you show up in algebra class!

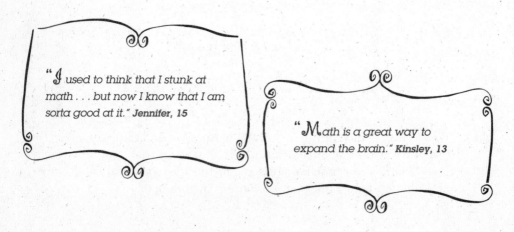

" *I* used to think that I stunk at math . . . but now I know that I am sorta good at it." **Jennifer, 15**

" *M* ath is a great way to expand the brain." **Kinsley, 13**

Lines and Their Slopes

Do you ski? I used to, but now I like to dance instead. Anyway, ski slopes on mountains can be gentle or steep, and the same is true for lines that we graph!

One way we can study lines is to talk about how steep they are. We call this "steepness" the *slope*. For the moment, we'll write slopes as *fractions*.

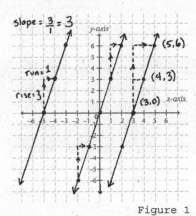

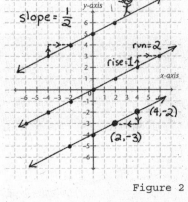

Figure 1   Figure 2

For positive slopes, the bigger the number is, the steeper the slope is. The slope of all three lines on this first graph is $\frac{3}{1}$. We can travel from one point to another point on the lines by "rising" up 3 units and "running" to the right 1 unit. The slope of all three lines on the second graph is $\frac{1}{2}$. In this case, notice how we can get from one point to another point on the lines by "rising" up 1 unit and "running" to the right 2 units. See a pattern? The *slope* of the line is defined as the ratio (or fraction) of *how far we rise* to *how far we run*. In other words:

$$\textbf{slope} = \frac{\textbf{rise}}{\textbf{run}}$$

"Ring Ring" What's It Called?

Slope

The **slope** of a line is the <u>steepness</u> of the line. It can be written as the ratio $\frac{rise}{run}$ where the "rise" represents vertical change (up or down), and the "run" represents horizontal change (right or left) between any two points on the line.

Notice that we call the slope a *ratio*.* That's because no matter which two points we travel between, we'll still end up with the same *ratio* of rise to run—the same slope! For example, in Figure 1 (p. 289), let's look at the line on the right. From the point (3, 0) to the point (4, 3), we rise up 3 units, and we run to the right 1 unit, right? So then $\frac{rise}{run} = \frac{3}{1}$, just as we would hope, because we already knew the slope was $\frac{3}{1}$.

But what if we traveled between two points farther apart, like from (3, 0) to (5, 6)? Look at the graph again. In this case, we'd rise up 6 units and go to the right 2 units, see that? Well, then, that $\frac{rise}{run}$ ratio is $\frac{6}{2}$. Just reduce the fraction and we get $\frac{3}{1}$, the same ratio—the *same slope*. In fact:

> **You can figure out the slope for *any graphed line* by counting the rise and the run between *any* two clearly marked points on the line, writing this as a slope fraction, and reducing it!**

Rise over Run . . . and the Bubblegum Trick

*P*ersonally, I always had a hard time remembering that the slope is $\frac{rise}{run}$ and not $\frac{run}{rise}$. If only there were a way to picture the correct order of the slope fraction in our heads! Oh, but there is . . .

Do this (or *imagine* doing this): Chew some gum. Two huge wads—like, a whole pack's worth of gum.[†] Yep, get it nice and soft and sticky. Then, stick one wad of gum on the *front* of a small dinner plate, and stick the other wad on the *back* of the plate. Now hold onto the *top piece* of gum, and let go of the dinner plate. If you were holding the plate high up in the air, you now have a broken dinner plate and haven't learned much about math (or gravity).

· · · · · · · · · ·

* To review ratios, see Chapter 16 in *Math Doesn't Suck*.
† To prevent choking, though, never chew more than one or two sticks *at a time*.

However, if you were holding the plate just a few inches above a table, then what you have is one long, stringy piece of gum going from your fingers down to the plate and one piece of flattened gum on the table under the plate.

If you pretend that the dinner plate is the line of a fraction, then in the top of the fraction, you have a tall, skinny, stringy piece of gum (the rise), and on the bottom of the fraction, you have a flattened piece of gum, extending out to the right and left (the run). And now we have the bubblegum dinner plate picture of this: $\frac{rise}{run}$. If you imagine the gum, you'll always remember which is which!

I guess I should have mentioned *not* to use a nice tablecloth for this little demonstration.

Negative slope

Slopes can be negative, too: These lines *slant in the other direction*. Check out the following graphs with negative slopes, and compare them to the graphs on p. 289, just to get a feeling for how they're different.

Look at the dotted lines I've drawn on the graphs below: With a negative slope, *either* the rise will be negative, *or* the run will be negative.

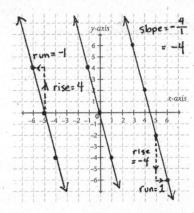

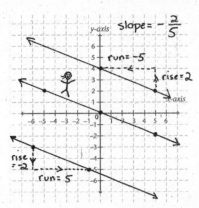

Notice on the above graphs that a negative "rise" just means we go *down* instead of up, and a negative "run" means we move to the *left* instead of the right. A negative slope will have a negative run *or* a negative rise, but not both.

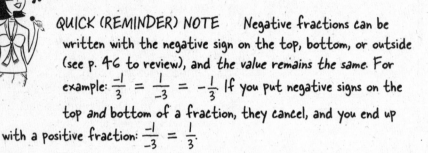

QUICK (REMINDER) NOTE   Negative fractions can be written with the negative sign on the top, bottom, or outside (see p. 46 to review), and *the value remains the same.* For example: $\frac{-1}{3} = \frac{1}{-3} = -\frac{1}{3}.$ If you put negative signs on the top *and* bottom of a fraction, they cancel, and you end up with a positive fraction: $\frac{-1}{-3} = \frac{1}{3}.$

On the graphs on p. 291, notice that my drawn dotted lines count the rise/run units *two ways*: once with a negative rise and positive run, and once with a positive rise and a negative run. Either way is fine! Because the negative slope can be written with the negative sign on either the top or the bottom, you'll end up with the *same slope* either way. After all, $\frac{-2}{5} = \frac{2}{-5}.$

By the way, for a line with a positive slope, you could choose to travel between two points by going down and to the left, if you really wanted to. Look at Figure 2 (p. 289). Between, say, the points $(4, -2)$ and $(2, -3)$, you could travel *down* 1 unit and to the *left* 2 units. If you did, technically you'd have to write the slope fraction as $\frac{-1}{-2}$, because you traveled in the negative direction both times, right? But, of course, this reduces to $\frac{1}{2}$; we still get the correct slope. You can't go wrong!

Step By Step

Identifying the slope from a graphed line:

Step 1. Pick any two points on the line whose (x, y) coordinates you know.

Step 2. Travel from one point to the other: First move straight up or down a certain number of units, counting them (this number will be the "rise"), and then move to the right or left, counting these units (this number is the "run"). Remember: Going *down* gives us a *negative rise*, and going to the *left* gives us a *negative run*.

Step 3. Create the slope fraction $\frac{rise}{run}$ and reduce it. That's the slope of the line!

QUICK NOTE   In Step 2, when you're counting rise and run units, you can draw your own dotted lines, like I have on p. 289 and p. 291, if it helps.

Watch Out!

When we plot points, we always count the *x*-direction first, because points look like this, (*x*, *y*), where the *x*-coordinate comes first. But when we count from one point to another point, to find the rise and run, we have to remember that the *rise* (the *y*-direction) goes on top of the fraction; that's sort of "first," right? The danger is that it's easy to start thinking that the *y*-coordinate comes first in the (*x*, *y*) points. (It's easier to confuse than you'd think!) So when you're working with points and slopes, keep the *mental image* that points are written alphabetically: (*x*, *y*), and the *mental image* of the bubblegum for the $\frac{rise}{run}$, and you'll keep them straight in your brain!

Doing the Math

For each line, name the coordinates of the points that have been marked, and then identify the slope of the line. I'll do the first one for you.

1. Line A

<u>Working out the solution:</u> Looking at the graph on the following page, we'll find the line marked A. Let's figure out what those points are. Hmm. The point on the left seems to be (-2, -1); remember that the x-coordinate always comes first, and that makes sense because x comes before y in the alphabet. The other marked point is (1, -5). Okay, now let's go from one to the other. It doesn't matter which one we start with, so let's start with (-2, -1). To go from here to (1, -5), we could move *down* (negative direction) 4 units—that's the rise—and to the *right* (positive direction) 3 units—that's the run. That means the slope $= \frac{rise}{run} = \frac{-4}{3}$, which can also be written as $-\frac{4}{3}$. Nothing to reduce, so we're done!

<u>Answer:</u> The points marked on line A are (-2, -1) and (1, -5), and its slope = $-\frac{4}{3}$.

2. Line B

3. Line C

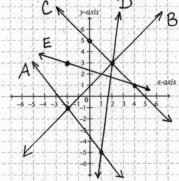

4. Line D

5. Line E

(Answers on p. 325)

You'll study lines in more depth once you get into algebra, but for now, I want to introduce you to the $y = mx + b$ form of writing *equations of lines*, so they're easier when you get there.

Slope-Intercept Form for Lines: $y = mx + b$

In this chapter so far, we've been working with *linear functions*: functions that create straight lines when you plot their points. Here's an interesting thing about functions that create straight lines: They can *always* be rewritten in the form $y = mx + b$, where m and b are both numbers. <u>So if you see a function that takes the form $y = mx + b$, you can bet it'll make a straight line!</u>

Remember on p. 289 how we compared the slope of a line to the steepness of a <u>mountain</u>? Well, the letter m stands for the "slope" in these equations! Yep, if an equation for a line is written in $y = mx + b$ form, then just look to see what m is, and that's the slope!

The b stands for the y-intercept—the spot on the <u>y-axis</u> where our line crosses it.

Here are some examples of equations in the form $y = mx + b$. Just fyi, this is actually the form we've been using all along!

$y = 3x + 1$
$y = x - 2$
$y = -\dfrac{1}{2}x + \dfrac{9}{2}$
$y = -\dfrac{2}{5}x$
$y = 5 - x$
$y = -x + 5*$

here, m= 3 and b= 1

here, m= 1 and b= -2

here, m= $-\frac{1}{2}$ and b= $\frac{9}{2}$

here, m= $-\frac{2}{5}$ and b=0
(the person walks on this line on p. 291)

here, m= -1 and b= 5
(these are the same line)

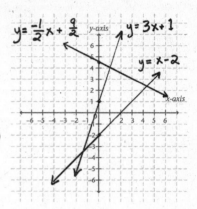

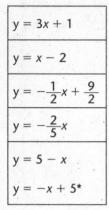

(the person walks on this line on p. 291)

"Ring Ring" What's It Called?

Linear Function

A **linear function** describes a line and can be written in the form $y = mx + b$, where m and b are both real numbers. (Yep, they could be zero, too.)

The m stands for the *slope*. Think of a ski slope; m stands for <u>m</u>ountain! The **y-intercept** is the number on the y-axis where the line crosses the y-axis. Its value is the b in $y = mx + b$. Imagine that the y-axis is a tall pole that you're practicing sword fighting against. Your <u>b</u>lade would strike the tall pole at the number b. Hiyaaah!

• • • • • • • • • •

* Do you see how these are two ways of writing the *same equation*? First take $y = 5 - x$ and rewrite the subtraction as "adding a negative" like we learned to do on p. 7. So: $y = 5 - x \rightarrow y = 5 + (-x) \rightarrow y = -x + 5$. Voilà! BTW, this is line C from p. 295.

Watch Out!

The *y*-intercept is the spot where the line crosses the y-axis, not the *x*-axis. Remember that a *tall* pole is what your <u>bl</u>ade is striking, so *b* is where the line crosses the *tall y*-axis!

QUICK NOTE   On p. 289, I told you that we'd write slopes as fractions "for the moment." We did that so it would be easier to see how the slope = $\frac{rise}{run}$. But if you have a slope of $\frac{4}{1}$, then it's customary to say that the slope = 4. Just remember that this means that the rise = 4 and the run = 1, and you'll be fine!

Look at the top line on p. 289—the one that the little person is walking on. Notice that it crosses the *y*-axis at the point (0, 5). Indeed, its *y*-intercept is 5, so *b* = 5. Also, we know that the slope is $\frac{1}{2}$, which means that $m = \frac{1}{2}$. So guess what? We can write out the equation of the line, can't we? Just filling in our numbers for *m* and *b* into $y = mx + b$, we get $y = \frac{1}{2}x + 5$. The middle line on that same graph has the same slope but a *y*-intercept of 0; do you see that? So the equation of that line could be written $y = \frac{1}{2}x$. I bet you can figure out what the equation for the bottom line would be!

Building Linear Equations

Let's say that we don't have a graph to look at, but I tell you that I'm thinking of a line, and all I say is that it has a slope of 4 and a *y*-intercept of 8. Can you tell me which line I'm thinking of? Well, yes, you can! Like we just did above, all you have to do is plug in 4 for *m* and 8 for *b* in the good 'ol equation $y = mx + b$, and you'll get $y = 4x + 8$. Nicely done!

Instead, if I tell you the slope is -1 and the y-intercept is -9, you could figure out that the line's equation must be $y = -1x + (-9)$; in other words, $y = -x - 9$, right?* Let's practice this, and brush up on our graphing, too.

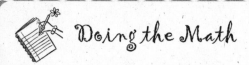 *Doing the Math*

For each of the following values of m and b:

a. Create its linear equation.

b. Graph its line by finding and plotting three points.

It's a lot to think about all at once, but I know you can do it! I'll do the first one for you.

1. $m = 0.5, b = -3$

<u>Working out the solution</u>: a. Plugging these values into $y = mx + b$, we get the linear equation $y = (0.5)x + (-3)$; in other words: $y = (0.5)x - 3$. So far, so good?
b. Now, to find points, just like on p. 284, let's pick some random values of x and find their y partners: If $x = 0$, then $y = (0.5)(0) - 3 \rightarrow y = -3$. First pair: $(0, -3)$. Next? Oh, let's see, how about $x = 2$,† so $y = (0.5)(2) - 3 \rightarrow y = 1 - 3 \rightarrow y = -2$. Next pair: $(2, -2)$. Hmm. Let's use $x = 6$, and we'll get $y = (0.5)(6) - 3 \rightarrow y = 3 - 3 \rightarrow y = 0$. Our third pair is: $(6, 0)$. Time to graph! Lookie there, sure enough, the y-intercept is at -3 on the y-axis, just like b said it would be. And if you count the rise and run, you'll find that, yep, $m = \frac{rise}{run} = \frac{1}{2} = 0.5$.

• • • • • • • • • •

* Remember that $-1 \cdot x = -x$. It's just a sneaky negative, that's all!
† I picked $x = 2$ on purpose to get rid of the decimal, because $0.5 = \frac{1}{2}$, and so $(0.5)(2) = 1$. With practice, you'll start to see how to pick "good" values of x that will get rid of decimals and fractions and make your life easier!

<u>Answer:</u> $y = (0.5)x - 3$ (and see graph below.)

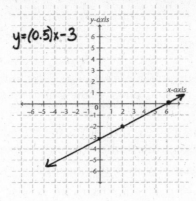

$y=(0.5)x-3$

2. $m = 2, b = 1$

3. $m = -2, b = -1$

4. $m = 0, b = 4$

5. $m = \frac{2}{3}, b = 0$ (Hint: Pick values of x that are divisible by 3 to make it easier!)

6. $m = -0.25, b = -1$ (Hint: Write the decimal as a fraction and pick x values divisible by 4.)

(Answers on p. 326)

QUICK NOTE   Equations for lines can be written in other forms besides $y = mx + b$. For example, $2y - 2x = -6$ and $(7x - 2) + (y + 3) = 5$ are both linear equations, too. You could graph them and everything! But rest assured that with a little algebraic rearranging, you can turn any linear equation into the form $y = mx + b$. (In fact, these two equations could be rewritten as $y = x - 3$ and $y = -7x + 4$, respectively.) You'll do more of this stuff later in algebra. I just wanted you to be aware of it!

Other Good Things to Know, Moving Forward

The Mutant Equations

As you encounter more and more $y = mx + b$ equations for lines, some might seem to be missing parts . . . these are what I like to call the "mutants." Let's check 'em out!

If there is no <u>b</u>, as in **y = 2x**, this means $b = 0$, and the line will cross through the origin (since that's where 0 is on the y-axis).

If there is no <u>m</u>, as in **y = x + 8**, this means $m = 1$.

If there is no <u>m or x</u>, as in **y = 3**, this means $m = 0$, and the line will be horizontal—totally flat. (See problem 4 on p. 299.) If you were skiing, this slope would be pretty boring.

If there is no <u>m or y</u>, as in **x = 4**, then this will be a *vertical* line. These vertical lines are said to have "no slope"—the slope's value is undefined.* If you were skiing, it would be a steep cliff, which wouldn't be boring to ski, it would be *impossible* to ski. Yikes!

Slope of Zero vs. No Slope

Saying "slope of zero" and "no slope" might sound the same at first, but they are very different! Lines with no slope don't get *any* number assigned to their m: Nope, not a positive, negative, or zero number. At least the zero slope *has* a value. Its value just happens to be 0, that's all!

Example of line with zero slope: The line **y = 3** has zero slope, because $m = 0$. That's why the mx term disappeared; the x got multiplied by zero and poof! All gone. Let's think about what $y = 3$ means. This equation doesn't care what x is. In fact, no matter what x is, y will *always* be equal to 3. Some points on this line would be $(-2, \mathbf{3})$, $(0, \mathbf{3})$, and $(82, \mathbf{3})$. When you graph it, you'll see that, yep, the line is totally flat—just what you'd expect a boring "zero" slope to be.

Example of line with no slope: On the other hand, the line **x = 4** has *no* slope; the slope is said to be "undefined." No matter what y is, x will always have to be 4. Some points on this line are $(\mathbf{4}, -13)$, $\left(\mathbf{4}, -\frac{2}{5}\right)$, and $(\mathbf{4}, 31)$. Try graphing it, and you'll see that it's a straight, vertical line!

• • • • • • • • • • •

* Vertical lines' slopes are so steep, it's like they have a slope of infinity! That's why we say these slopes are *undefined*.

What's the Deal?
Functions vs. Equations

I started out this chapter talking about *functions*, and how to graph points and lines from them. So why, later in the chapter, did I start using the much broader term *equation* instead of *function*? Well, even though most equations for lines are functions, *vertical lines are NOT functions!*

This is a technical definition thing. You see, on p. 267, we defined a *function* as a type of equation and mentioned that for every ingredient, a function delivers only *one kind* of sausage. With functions, each x can result in only one y—not *more* than one y. In vertical lines like $x = 3$, however, there are many different sausages (y's) that would each get produced from that one x ingredient, 3. This is impossible for a sausage factory! You put in raspberries one day and get one result; then, the next day, you put in raspberries and get something else? Nope, not possible. Sausage factories are very consistent: They always do the *same thing* to the ingredients, whether you stick in raspberries or the number 3. So a group of points like (3, −4), (3, 29), (3, 163) could never have come out of a sausage factory. (One x can never result in more than one y in functions.) That's why vertical lines like $x = 3$ are not considered to be functions.

On the other hand, you *could* perhaps substitute raspberries for strawberries, and maybe no one could tell the difference between the two resulting sausages. You're just being clever with your recipe, that's all. (More than one x *can* deliver the same y in functions.) So a horizontal line, like $y = 29$ with points like (3, 29), (4, 29), and (5, 29), *is* considered a function. See the difference?

Besides vertical lines, there are all sorts of other (nonlinear) equations you could graph that *aren't* functions, but you don't need to worry about them right now.*

Also, you might not need to understand this formal distinction between "function" and "equation" until algebra, but some teachers do talk about it in pre-algebra, so I wanted you to be covered. Besides, they don't usually explain it with sausage recipes, and I thought you might like that.

. . . .

* In case you're curious, here's an example of another kind of equation that is *not* a function: $x^2 + y^2 = 1$. Much later in algebra, you'll see that this graph looks like a circle with a radius of 1!

Takeaway Tips

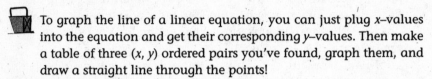 Ordered pairs of numbers, in the form (x, y), can be graphed as single points on the two-dimensional coordinate plane.

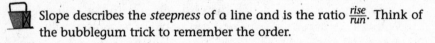 To graph the line of a linear equation, you can just plug x–values into the equation and get their corresponding y–values. Then make a table of three (x, y) ordered pairs you've found, graph them, and draw a straight line through the points!

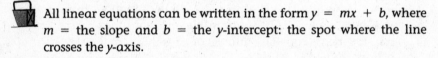 Slope describes the *steepness* of a line and is the ratio $\frac{rise}{run}$. Think of the bubblegum trick to remember the order.

All linear equations can be written in the form $y = mx + b$, where m = the slope and b = the y-intercept: the spot where the line crosses the y-axis.

A Final Word

I'm very proud of you—some of the stuff in this book ain't easy. But actually, that's good news . . . let me explain.

Whenever you don't understand how to do a math problem, this is actually a good thing, because now you have an opportunity to exercise the part of your brain that makes you stronger, more capable, and successful in life: the part that *does not give up*.

When you're struggling with something but you believe in yourself and you keep trying until you succeed, you not only become stronger, but also much more powerful. Doing math has a funny way of expanding our brains, making us better problem solvers, and strengthening our mental fortitude and stamina. As hard as math can seem sometimes, you're actually benefiting from it in ways you might not realize.

As you keep growing and developing into a young woman, you'll start to notice that there are some people who keep trying and others who give up. It might seem "cool" to give up and not care, but believe me, in a few short years, you'll either have exercised that "not-giving-up" muscle or you won't have. And this, right now, is the time that you are deciding what kind of woman you'll be, whether you realize it or not.

Seek out the things you don't understand, and seize opportunities to learn how to *think* in entirely new ways. Believe me, math will keep giving you these opportunities, so take them!

And guess what? Along the way, while you're getting mentally stronger and more able to conquer *any* challenge in front of you, math or otherwise, you'll also be learning more *math*—something else that will set you apart from the crowd and give you advantages as an adult that you can't even imagine.

So go get 'em, and use math to get a leg up and fulfill your dreams. Later on, you'll want to kiss yourself . . . *and* your math.

Math Test Survival Guide!

$\mathcal{M}$ath tests getting you down? Now you can relax! Because in the next few pages, you'll find everything you need to become a confident test-taker. (By the way, if you've recently failed a test, be sure to read pp. 59–61!)

What follows are tips and tricks that I developed throughout high school and even as a math major in college, taking *four* math classes at a time. I can tell you—they work!

Ahead of Time—Everyday Tips to Make Tests Easier!

1. Every night, rewrite your notes that you take in class, that *same* night. Just rewrite them. Easy, right? The stuff will stick in your brain way better, and if you wrote something down that doesn't make sense, you can easily ask your teacher about it the next day. This is much better than waiting until the night before a test to read your notes and then being confused! The simple act of rewriting your notes each night will also make your notebook look neater (try using colored pens!), and tonight's math homework easier and faster.

2. If your teacher collects your homework and doesn't usually hand it back until after the test, I highly recommend writing down one or two problems from each topic/assignment on a special piece of paper called "Example Problems" to keep in your notebook. I also recommend making scratch notes* while you do your homework, also to keep in your notebook. This is you, studying for a math test: "Didn't I used to understand this stuff? How the heck did I do these problems before? Hmm. I'll just check my handy-dandy 'Example Problems' and scratch notes!"

When a Test Is Announced

Say good-bye to those feelings of dread and those "aimless" hours of studying the wrong stuff. Say hello to calm, confident, time-saving test preparation! Here's the whole strategy leading up to the test:

1. Read your notes that you've rewritten so neatly.
2. Make a Pretend Cheat Sheet.
3. Make a Quickie Sheet.

(More on 2 and 3 below.)

Sure, you should do some practice problems, too. But you know how it can be hard to know *which* problems to practice? Now it'll be easy.

The Pretend Cheat Sheet

Once the test is announced and you know which topics will be covered, you don't have to start studying right away, but I do recommend that you start making your personalized *Pretend Cheat Sheet* right away.

Here's how: Read through your notes slowly, just to remind yourself which stuff is easy for you and which is harder. While you're doing this, take out a separate piece of paper and start the Pretend Cheat Sheet: This will be a *single* page, front and back, of everything you wish you could take with you into the test.

. . . .

* Basically, this is just doing your homework on a "messy paper," showing all your thoughts and all the steps, and then rewriting it neater for the homework that you turn in. For more details on this, see p. 275 in the Troubleshooting Guide of *Math Doesn't Suck*.

Pretend that you could take it with you to the test: What would you include? You might say, "I'd bring all my notes and my textbook, too!" But if you really could bring that stuff to the test, just think of how long it would take to find anything. That's why it's important to use just a *single page*.

Don't write down the easy stuff you already know; that would be a waste of precious space, right? Only write down the things (formulas, definitions, problem-solving strategies, things you tend to mix up, etc.) that you really wish you could bring with you into the test.

Check your book and homework to see if there are tidbits you want to include from there, too. Creating these sheets can be done with friends in a study group, but each person's page will probably end up looking a little different.

Here's a tip: Don't write so small that you can't read your writing! In fact, the more spaced out it is, the better. And since you only have one page, don't be surprised if you end up rewriting your Pretend Cheat Sheet a few times. You want to make sure you can fit everything you *really* want on it. When I was in high school and college, I would usually do a rough draft of my Pretend Cheat Sheet first, which would always be too long, and then I'd edit it down to the real deal.

While making the sheet, you'll also be reminded of things you still don't understand. However, since you're working on it so far in advance, you'll have plenty of time to either look in your textbook for the strategy, search a book like this or *Math Doesn't Suck* for answers, or ask a teacher for help.

At this stage, you can also do practice problems along the way—but only if you have time. Your main goal is to make a fabulous Pretend Cheat Sheet! (There's an example on the following page.)

4 to 5 Days Before the Test

Update your Pretend Cheat Sheet, especially if after you made it you learned new things that might be on the test. Also, now's the time to do practice problems from the sections that most confuse you. Keep referring to your personalized Pretend Cheat Sheet for guidance on what *you* should study! That way, you won't waste time studying stuff you already know cold. Just do a little each night.

Danica's Pretend Cheat Sheet for...

♡ Integers and Number Properties ♡

Order of Operations:
Pandas Eat Mustard on Dumplings and Apples with Spice
M & D together, A & S together: Left to Right!

Integers: Two negs → pos: $3-(-5) = 3+5$!
multiplication by -1 = "the opposite of"
✱ Always change subtraction → adding negs ✱
To mult./div. Integers: <u>COUNT</u> the neg signs
odd # → neg.   even # → pos.
✱ make sure it's a (single term)✱
For example: Don't count yet   Ready for counting
$$\frac{-5-2}{-2(-3)} \implies \frac{-7}{-2(-3)}$$

Associative Property: Parentheses move, like popularity
(Rule) → Rule: $(a+b)+c = a+(b+c)$ and $(ab)c = a(bc)$
Example: $(-2+3)+4 = -2+(3+4)$ and $(-2\cdot3)\cdot4 = -2\cdot(3\cdot4)$

Commutative Property: Numbers move, like cars' commute
(Rule) → Rule: $a+b = b+a$ and $ab = ba$
Example: $-2+3 = 3+(-2)$ and $(-2)(3) = (3)(-2)$

✱ Assoc. and Comm. properties work when there is <u>ONLY</u>
addition or <u>ONLY</u> multiplication, but neg. numbers are OK.

The Evening Before the Test

Do your math studying while you're fresh, before working on other
subjects. Skim through your book and notes as much as you want, but
definitely do this: <u>Read your Pretend Cheat Sheet five full times</u>, slowly,
absorbing as you go. And do a few more practice problems, including

one or two from each section that *used* to stump you but that you now understand. It's a good confidence booster, and it reinforces what you know!

Once you've finished that, it's time to make the Quickie Sheet! (See p. 311 for how to make it.) You'll pick three things from your Practice Cheat Sheet and write them on the Quickie Sheet—just the trickiest stuff like confusing formulas or rules. You'll look at this Quickie Sheet a few minutes before the test. You are *guaranteed* to remember these three things while you're taking the test, so choose them wisely. (I *still* use Quickie Sheets for live TV interviews. I write down three things I want to talk about. It keeps me focused and feeling confident.) You can either download and print a Quickie Sheet from kissmymath.com or copy the one on p. 311.

Now eat a healthy dinner, do some stretching, and get at least eight hours of sleep!

The Morning of the Test

Good morning! Eat a good breakfast. While you do, and even on your way to school, read over your Pretend Cheat Sheet and your Quickie Sheet. Read them calmly and slowly. Think nice, happy thoughts like "I can do this. It's not so bad."

Five Minutes Before the Test

Time to review the Quickie Sheet. Don't even THINK about reading any other math. It's too much to absorb! It's time to relax your mind and body, and release tension. It'll make a huge difference. (Also, make sure to put away all materials completely before the test starts, so the teacher doesn't think you're cheating by mistake!)

When the Test Is Handed to You

1. Don't read the test yet. *While your Quickie Sheet is still in your mind,* flip the test over to the blank side, and write down your own mini review sheet! (It's not cheating if you do it without help, after all!) You may want to jot down whatever was on your Quickie Sheet and perhaps a few other things. Don't spend long on this—just a minute or two. And it doesn't have to be neat. It's just for your reference.

2. Now, flip the test back over, and skim the *entire test*. See which problems look easiest to you and lightly circle them.

3. Do whichever problems you *feel* like doing first, referring to the notes on the back of the test as much as you want. After all, #3 might be something from your Quickie Sheet, and #1 might look really hard!

4. If you ever start to panic during the test, just take a moment, breathe deeply, and think about the calming words of encouragement on the Quickie Sheet. Relax your mind and your muscles. It makes all the difference. Now go get 'em!

By the way, save your Pretend Cheat Sheets. They come in very handy when it's time to study for final exams and sometimes even for the following year, believe it or not!

The Pressure of Time?

Once you've started practicing calming your mind before tests, if you find that the timed aspect of test-taking is still making you crazy, try this. Ask your teacher, "How many of tonight's homework problems should I be able to do in 30 minutes?" Then set a timer while you do your homework. Use your imagination when you start your homework: Imagine being in your classroom, the teacher saying "begin," etc. You can even look at a Quickie Sheet before you start. Really, pretend it's a test!

Using a timer for homework may be stressful at first, but this practice will help you focus your mind and learn how to remain calm under pressure. I know from experience. My mom taught me to do this in junior high! And the more often you do this, the better you'll get at it.

Remember, taking tests is a *skill* that is totally independent of the subject matter. It's not a bad skill to develop, lemme tell ya—you'll be taking tests for years to come. Why not get *good* at it? It'll make you a more powerful young woman, that's for sure!

Quickie Sheet

Hello there! I know you're about to take a math test. But you can let go of the math for the moment and review this sheet . . . *slowly*.

Relax your body.

Find tension spots, and imagine them melting away as you breathe deeply and calmly. Close your eyes and stretch your body a little as you breathe. Do this until you *feel better*.

"I can do this. I'm ready. Bring it on!"

Now continue taking slow, deep breaths, and confidently repeat this phrase to yourself, several times in a row, until you actually start to believe it. (You might not at first, and that's okay!) Closing your eyes can help with this, too. And you *can't* overdo this one. More is better!

Three Reminders

These can be math formulas, rules, tips—whatever will help you the most on the test. Now, read them while continuing to breathe calmly, and stay focused on that good, confident, relaxed feeling as much as you can.

    1. _____

    2. _____

    3. _____

Whenever you start to lose that good, confident, relaxed feeling, use the above tips to calm your mind and body. Reviewing a whole bunch of math before a test is fine, but trust me, for the last five minutes, this sheet will help you *much* more.

Finally . . . *smile!*
You're going to do GREAT!

Appendix

More Number Properties

Here are a few Number Properties, in addition to the ones I covered in Chapter 2 and Chapter 10, just in case you're forced to memorize their names. The properties themselves are pretty obvious. Seriously, this is stuff we learned in kindergarten but with fancier names!

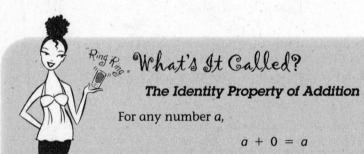

"Ring Ring" What's It Called?

The Identity Property of Addition

For any number a,

$$a + 0 = a$$

In other words, adding zero doesn't change a number's value. So, 0 is sometimes called the **additive identity** because a gets to keep its own *identity* when 0 is *added* to it!

The Identity Property of Multiplication

For any number a,

$$a \times 1 = a$$

In other words, multiplying by 1 doesn't change a number's identity, and that's why 1 is sometimes called the **multiplicative identity**.

In case you have to memorize these for class, just remember that the first two are called the "identity" properties because, hey—no identity crisis necessary here. Every number gets to stay who it always was!

Sets of Numbers

Counting Numbers*: This is the set of all numbers we use to count, starting with 1: {**1, 2, 3, 4, 5** . . . }. This set does not include any fractions or decimals of any kind.

Whole Numbers: This is just the counting numbers with zero added. I know, I know—why the "whole" new name? Who knows, but here's the set: {**0, 1, 2, 3, 4** . . . }.

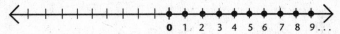

Integers: This is the set of all whole numbers and their opposites. So it extends in both directions forever, positive and negative. Also, it does not include any fractions or decimals. Check it out: {. . . **−5, −4, −3, −2, −1, 0, 1, 2, 3, 4, 5,** . . . }. On a graph, this set of numbers is represented by dots at each integer. Imagine them going on forever in both directions.

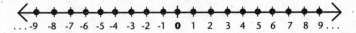

This book covers integers (aka "mint-egers") in Chapters 1 and 3.

Rational Numbers: This is the set of all logical, level-headed numbers. Just kidding. It's the set of all values that can be expressed as a *ratio* of two integers—in other words, expressed as a fraction of two integers. All fractions, terminating decimals, and repeating decimals are rational

* These are sometimes called *Natural Numbers*.

numbers, because they can be expressed as fractions. Examples of rational numbers are 0.5, -1, $\frac{163}{1}$, $0.\overline{8}$, and $-\frac{4}{99}$. To learn how to convert repeating decimals into fractions, see Chapter 12 in *Math Doesn't Suck*. It's impossible to represent the set of all rational numbers on a number line. There are an infinite number of them even within the segment from 0 to 1, with *irrationals* in between! (See p. 315 for more on this.)

Irrational Numbers: Likewise, this is not the set of crazy numbers; it's the set of all values that *cannot* be expressed as a ratio of integers. Examples of irrational numbers are π, $\sqrt{2}$, and 0.123456789101112 . . . For reasons you'll learn about later in math, there are way more irrational numbers than rationals. It's also impossible to represent the set of all irrational numbers on a number line. There are an infinite number of them even within the segment from 0 to 1 with *rationals* in between! (See p. 315 for more on this.)

Real Numbers: This is the set of all rational *and* irrational numbers. When you see "any number *a*" like on p. 312 and p. 313 in the definition boxes, it means that *a* can be any *real* number.

It's easy to represent the set of all real numbers on the number line: Just draw a solid line, extending in both directions, forever!

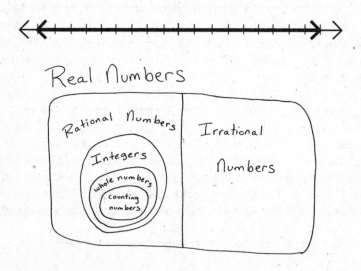

Here's a Venn diagram of these sets of numbers. As you can see, all counting numbers are whole numbers, and all whole numbers are integers. Also, if you're an integer, you can certainly *be written as a fraction*, so this means that all integers are considered rational numbers. Notice, too, that rationals and irrationals are mutually exclusive. A number has to be in one set or the other; it can't be in both.

There Are Infinitely Many Rational Numbers Between 0 and 1

Don't believe me? What if I listed some of them for you? I'll construct an infinitely long list of rational numbers between 0 and 1. How about $\frac{1}{1}$, $\frac{1}{2}$, $\frac{1}{3}$, $\frac{1}{4}$, $\frac{1}{5}$ . . . Do you see how, as this list continues, the numbers get smaller and smaller but never reach zero? The hundredth number on this list would be $\frac{1}{100}$, and the billionth number on the list would be $\frac{1}{1,000,000,000}$. *Very* close to zero . . .

This list will go on forever—the numbers will continue to get closer and closer to zero, but *never reach zero*. So there really are infinitely many rational numbers between 0 and 1!

Can you construct a different, infinitely long list of rational numbers between 0 and 1? In fact, a list that doesn't include any of the same fractions as the ones in the list above?*

Of course, there are also infinitely many rational numbers in *every other* segment of the number line, not just 0 to 1. Wanna create an infinite list of fractions in the segment between 8 and 9? Just *add* 8 to each number in our above list. For example, instead of $\frac{1}{1}$, $\frac{1}{2}$, $\frac{1}{3}$, $\frac{1}{4}$, $\frac{1}{5}$ . . . we could write: 9, $8\frac{1}{2}$, $8\frac{1}{3}$, $8\frac{1}{4}$, $8\frac{1}{5}$ . . . Yep, these numbers will always be between 8 and 9, getting closer and closer to 8, but never actually reaching it.

There Are Infinitely Many *Irrational* Numbers Between 0 and 1

We could also make an infinitely long list of *irrational* numbers between 0 and 1. This is a little tougher, but let's do it anyway! Let's see, a pretty famous irrational number is π, whose value is approximately 3.14. One irrational number in the segment between 0 and 1 would be $\frac{\pi}{4}$. After all, it's just a tiny bit bigger than $\frac{3}{4}$, right?

Check out this list, $\frac{\pi}{4}$, $\frac{\pi}{5}$, $\frac{\pi}{6}$, $\frac{\pi}{7}$, $\frac{\pi}{8}$ . . . Do you see how the fractions keep getting smaller and smaller, but never reach zero? For example, in this list, the hundredth term is $\frac{\pi}{103}$ (remember that the first term has a 4 on the bottom not a 1), and the billionth term in this list would be: $\frac{\pi}{1,000,000,003}$. (To help conceive the values of these irrational numbers, imagine a 3 wherever you see π, and you'll be very close.)

. . . .

* Email them to me at danica@kissmymath.com!

Each of these terms is irrational, they each lie on the number line between 0 and 1, and there are infinitely many of them—so we've shown that there are indeed infinitely many irrational numbers between 0 and 1. Reread this section a few times if you need to; it's pretty advanced stuff!

Once you've digested this, can you construct an infinite list of irrational numbers that all live on the number line between 5 and 6? How about between −6 and −5?*

By the way, making infinite lists of numbers like this is part of what you'll do in pre-calculus someday. I don't know why, but I've always liked stuff involving infinity.

While We're Talking About Values: a Small Chat About Fractions, Decimals, and Percents

Think about percents for a moment. If I say "fifty percent," it doesn't *mean* anything unless we say *what* it's 50% of, right? In other words, it doesn't represent a value; it only represents a "portion" of something.

On the other hand, if I say "one half," I *might* be talking about $\frac{1}{2}$ as a portion of something else (for example: $\frac{1}{2}$ of $20), *or* I might mean $\frac{1}{2}$ as a value, which is a rational number that lives on the number line.

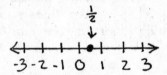

In fact, since fractions can be interpreted in both ways—as portions of another value *or* as a value in their own right—we can end up with expressions like this:

$$\frac{3}{4} \text{ of } 4 \qquad \text{three } \frac{2}{3}\text{'s} \qquad \frac{1}{2} \text{ of } \frac{2}{3} \qquad \frac{2}{3} \text{ of } \frac{3}{4}$$

Think about what each of these is saying. Can you evaluate them? If you're having trouble solving these, read Chapter 15 in *Math Doesn't Suck*. After you've attempted them, here are the answers.[†]

Moving from pre-algebra into algebra it'll be very helpful for you to remember that fractions—and decimals—can be thought of *both* ways: as portions of something else and as values in their own right. Every single fraction and decimal (yes, even the irrational ones) has a *value* that can theoretically be graphed on the number line.

- - - - - - - - - -

* Email them to me at danica@kissmymath.com!
† 3, 2, $\frac{1}{3}$, and $\frac{1}{2}$

Powers: Ms. Exponent at Work

You're probably already familiar with *squares* from your multiplication facts: $2 \times 2 = 4$, $3 \times 3 = 9$, $4 \times 4 = 16$, etc. Following is a table of some other commonly used powers. It will really help you on tests and in homework to learn the ones that are in bold. They'll come up a lot! I included the other ones in case you were curious about the patterns that can happen. Look at the last two digits in the row of the powers of 5; no matter what the exponent is, they all end in 25. Can you figure out why that happens?

squared	cubed	4th power	5th power	6th power	7th power	8th power
$2^2 = 4$	$2^3 = 8$	$2^4 = 16$	$2^5 = 32$	$2^6 = 64$	$2^7 = 128$	$2^8 = 256$
$3^2 = 9$	$3^3 = 27$	$3^4 = 81$	$3^5 = 243$	$3^6 = 729$	$3^7 = 2{,}187$	$3^8 = 6{,}561$
$4^2 = 16$	$4^3 = 64$	$4^4 = 256$	$4^5 = 1{,}024$	$4^6 = 4{,}096$	$4^7 = 16{,}384$	$4^8 = 65{,}536$
$5^2 = 25$	$5^3 = 125$	$5^4 = 625$	$5^5 = 3{,}125$	$5^6 = 15{,}625$	$5^7 = 78{,}125$	$5^8 = 390{,}625$
$6^2 = 36$	$6^3 = 216$	$6^4 = 1{,}296$	$6^5 = 7{,}776$	$6^6 = 46{,}656$	$6^7 = 279{,}936$	$6^8 = 1{,}679{,}616$
$7^2 = 49$	$7^3 = 343$	$7^4 = 2{,}401$	$7^5 = 16{,}807$	$7^6 = 117{,}649$	$7^7 = 823{,}543$	$7^8 = 5{,}764{,}801$
$8^2 = 64$	$8^3 = 512$	$8^4 = 4{,}096$	$8^5 = 32{,}768$	$8^6 = 262{,}144$	$8^7 = 2{,}097{,}152$	$8^8 = 16{,}777{,}216$
$9^2 = 81$	$9^3 = 729$	$9^4 = 6{,}561$	$9^5 = 59{,}049$	$9^6 = 531{,}441$	$9^7 = 4{,}782{,}969$	$9^8 = 43{,}046{,}721$
$10^2 = 100$	$10^3 = 1{,}000$	$10^4 = 10{,}000$	$10^5 = 100{,}000$	$10^6 = 1{,}000{,}000$	$10^7 = 10{,}000{,}000$	$10^8 = 100{,}000{,}000$

Isn't it amazing how quickly numbers can get big with exponents? For example, how crazy is it that 6^4 equals something as big as 1,296? Or that $9^7 = 4{,}782{,}969$? To me, it's sort of incredible; but then again, we know how powerful those high-ranking executives in their top-floor offices can be. Yep, Ms. Exponent does it again. Well, *I* was impressed, okay?

Answer Key

For the fully explained solutions, visit the "Solution Guides" page at kissmymath.com.

DTM Chapter 1, p. 4

2. −12, −5, 0, 3

3. −10, −7, −4, 6

4. −8, −1, 2, 7

DTM Chapter 1, p. 9

2. 2

3. −4

4. −12

DTM Chapter 1, p. 12

2. 6

3. 4

4. 2

5. 0

DTM Chapter 1, pp. 16–7

2. 2

3. −13.6

4. −11

5. 0

DTM Chapter 2, p. 27

2. 58

3. 90

4. 8

DTM Chapter 2, p. 28

2. **a**. yes; **b.** 10

3. **a**. nope, not equal; **b.** 8 and 0

4. **a**. yes; **b.** 15

5. **a**. nope, not equal; **b.** 12 and 24

6. **a**. nope, not equal; **b.** 2 and 8

DTM Chapter 2, pp. 33–4

2. $y + (-10)$ or $y - 10$
3. not allowed
4. z

DTM Chapter 3, pp. 47–8

2. 7
3. $\frac{8}{3}$ or $2\frac{2}{3}$
4. $-\frac{1}{3}$
5. 0
6. 2

DTM Chapter 4, pp. 57–8

2. 7
3. 2
4. $\frac{1}{2}$
5. $\frac{5}{2}$ or $2\frac{1}{2}$

DTM Chapter 5, pp. 69–70

2. mean = 2.5,
   median = 2.5,
   there is no mode
3. mean = 3, median = 3,
   mode = 3
4. mean = 6, median = 5,
   modes = 2 and 7
5. mean = -2,
   median = -1.5,
   modes = -4 and -1

DTM Chapter 6, pp. 86–7

2. $4 + 3$ 🌸, 7, 1, 5, and 4.6
3. 8, 7, -8, and 2 ☺ $+ \frac{6}{☺}$

DTM Chapter 6, p. 96

2. Two terms total; the variable is z; the coefficient is -4; the constant is 7.

3. Two terms total; the variables are n and m; the coefficients are 1 and -1; no constants.

4. Four terms total; the variables are a, b, and c; the coefficients are 1, -5, and $\frac{2}{3}$; the constant is 0.2.

5. There are three terms total; the variables are x and y; their coefficients are $\frac{3}{5}$ and -1; the constant is -9.

DTM Chapter 7, pp. 105

2. $7j$
3. $14c$
4. $0.2y$
5. $\frac{1}{4}z$
6. $6t + 10$

DTM Chapter 8, pp. 112–3

2. $-16g^2h$
3. $5a^2b$

4. $2w^2$

5. 0

DTM Chapter 8, p. 119

2. $2xy + z$

3. $-\dfrac{2}{b}$

4. -6

DTM Chapter 9, pp. 126–7

2. $5 + 1g + 1h$ or
$5 + g + h$

3. $4a + 10b + (-6b^2)$ or
$4a + 10b - 6b^2$

4. $-3xy$

DTM Chapter 10, pp. 133–4

2. 19

3. 32

4. 79

DTM Chapter 10, p. 140

2. $9 + (-h)$; or $9 - h$

3. $10 + (-3xy) + 12y$ or
$10 - 3xy + 12y$

4. -8

5. $7ab + (-3a) + 1$; or
$7ab - 3a + 1$

DTM Chapter 11, p. 149

2. a. Yes, this is a math sentence; it's an equation. **b.** This can be translated in several ways; here are some: Two times x, minus one, <u>equals</u> zero. Twice x minus 1, <u>equals</u> zero. One less than twice x <u>is</u> equal to zero. One less than two times x <u>is</u> equal to zero.

3. a. No, it is not a math sentence; it's just an expression. **b.** One third of y, plus three, plus x

4. a. Yes, this is a math sentence; it's an inequality. **b.** a <u>is</u> greater than or equal to two.

5. a. No, this is not a math sentence; it's just an expression. **b.** g plus zero.

6. a. Yes, this is a math sentence; it's an inequality. **b.** One third of z <u>is</u> less than seven.

DTM Chapter 11, pp. 150–1

2. a. This is an inequality.
b. $7 < 2x$

3. a. This is an inequality.
b. $13 > 3c$

4. a. This is an expression.
b. $3c + 12$

5. a. This is an expression.
b. $\dfrac{y}{2} - 5$

6. **a.** This is an inequality.
   b. $7 > \frac{w}{4}$

7. **a.** This is an equation.
   b. $\frac{x}{3} + 8 = 11$

DTM Chapter 11, pp. 154–6

2. She has $2s - 5$ dollars now.

3. $f = \frac{1}{4} \times \frac{4}{5}$ or $f = \left(\frac{1}{4}\right)\left(\frac{4}{5}\right)$.
   BTW, $\frac{1}{4}$ of $\frac{4}{5} = \frac{1}{5}$.

4. Each friend got $\frac{s - 5}{6}$ frozen grapes.

5. Chris has $\frac{s - 10}{2}$ text messages now.

6. Sarah has $2y + 9$ ringtones now.

7. After Suzanne goes home, there are $\frac{1}{5}m - 2$ puppies left at the store. (If you don't simplify, it looks like this: $m - \frac{4}{5}m - 2$.)

DTM Chapter 12, pp. 170–1

2. $4(x + 3)$, and to unwrap it, divide by 4 and then subtract 3.

3. $4y + 3$, and to unwrap it, subtract 3 and then divide by 4.

4. $\frac{z + 3}{4}$, and to unwrap it, first we'd have to multiply by 4 and then subtract 3.

5. $5\left(\frac{w}{2} - 1\right)$, and to unwrap it, first divide by 5, then add 1, and then multiply by 2.

6. $\frac{6n - 5}{7}$, and to unwrap it, we'd first multiply by 7, then add 5, and then divide by 6.

DTM Chapter 12, p. 179

2. $x = -3$

3. $x = 6$

4. $x = 8$

5. $x = 2$

DTM Chapter 12, p. 188

2. $x = 1$

3. $x = -6$

4. Equation is true for *all* values of x.

5. $x = 3$

6. $x = 8$

DTM Chapter 13, pp. 193–4

2. Trudy started with $100.

3. Brittany started with 47 frozen grapes.

4. Chris started with 26 text messages.

5. Sarah had 21 ringtones yesterday.

6. There were 10 puppies at the store that morning.

DTM Chapter 13, pp. 203–4

2. The purse costs $20. (No, it doesn't cost $40.)

3. **a.** Mom's offer is better
b. 20 hours

4. Leslie is 10, Duncan is 14, and Hunter is 17.

DTM Chapter 14, p. 214

2.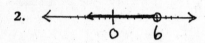

3.

4.

DTM Chapter 14, p. 223–4

2. $x < 8$

3. $x > -8$

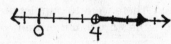

4. $x \leq -2$

5. $x > 4$

DTM Chapter 15, p. 241

2. $2^5 \times 5^2$

3. 10^6

4. $12^4 \times 7^3$

5. $(0.2)^4$

DTM Chapter 15, p. 248

2. -25

3. -150

4. -145

5. 0

DTM Chapter 15, pp. 251–2

2. $-\dfrac{1}{8}$

3. 0

4. -8

5. -12

DTM Chapter 16, p. 258

2. $\dfrac{81}{256} x^4$

3. can't distribute

4. $-\dfrac{81}{256} x^4 y^4$

5. $\dfrac{81}{256} x^4$

2.

X → f(x)	(x, f(x))
−6 → −14	(−6, −14)
−3 → −8	(−3, −8)
0 → −2	(0, −2)
3 → 4	(3, 4)
9 → 16	(9, 16)

3.

X → f(x)	(x, f(x))
−6 → −7	(−6, −7)
−3 → −5	(−3, −5)
0 → −3	(0, −3)
3 → −1	(3, −1)
9 → 3	(9, 3)

4.

X → f(x)	(x, f(x))
−6 → 42	(−6, 42)
−3 → 12	(−3, 12)
0 → 0	(0, 0)
3 → 6	(3, 6)
9 → 72	(9, 72)

2.

Ingredient → Sausage

X → y	(x, y)
−4 → −3	(−4, −3)
−1 → 0	(−1, 0)
0 → 1	(0, 1)
3 → 4	(3, 4)

3. Ingredient → Sausage

X → y	(x, y)
−4 → 10	(−4, 10)
−1 → 7	(−1, 7)
0 → 6	(0, 6)
3 → 3	(3, 3)

4. Ingredient → Sausage

X → y	(x, y)
−4 → −21	(−4, −21)
−1 → −9	(−1, −9)
0 → −5	(0, −5)
3 → 7	(3, 7)

2.

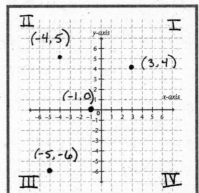

3.

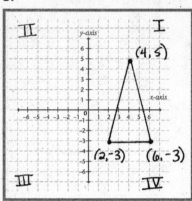

4.

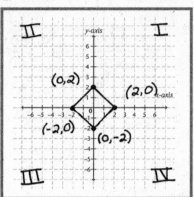

DTM Chapter 18, pp. 284–5

2.   **a.**

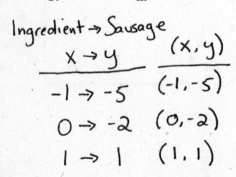

Ingredient → Sausage

$x \to y$	(x, y)
$-3 \to 0$	$(-3, 0)$
$0 \to 3$	$(0, 3)$
$3 \to 6$	$(3, 6)$

b.

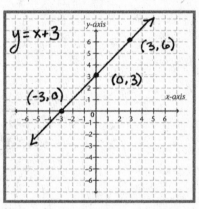

$y = x + 3$

3.   **a.**

Ingredient → Sausage

$x \to y$	(x, y)
$-1 \to -5$	$(-1, -5)$
$0 \to -2$	$(0, -2)$
$1 \to 1$	$(1, 1)$

b.

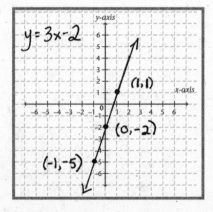

$y = 3x - 2$

4.   **a.**

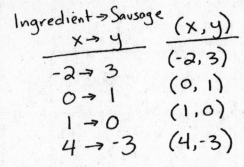

Ingredient → Sausage (x, y)

$x → y$

$-2 → 3$	$(-2, 3)$
$0 → 1$	$(0, 1)$
$1 → 0$	$(1, 0)$
$4 → -3$	$(4, -3)$

b.

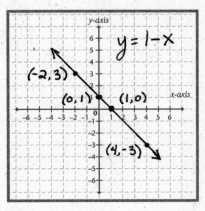

5.   **a.**

Regardless of the points you pick, the *line* you draw should be this same line. I've marked the points $(-5, -2)$, $(-1, 2)$, and $(1, 4)$, but your points will probably be different.

b.

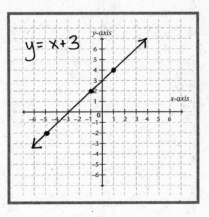

DTM Chapter 18, pp. 294–5

2. The points are $(-2, -1)$ and $(2, 3)$, and the slope of the line is 1.

3. The marked points are $(0, 5)$, $(2, 3)$, and $(4, 1)$. The slope of the line is -1.

4. The marked points are $(1, -5)$ and $(2, 3)$, and the slope of the line is 8.

5. The marked points are $(-2, 3)$ and $(4, 1)$, and the slope of the line is $-\frac{1}{3}$.

2. a. $y = 2x + 1$
   b.

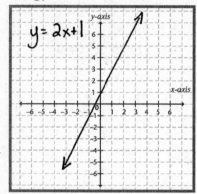

3. a. $y = -2x - 1$
   b.

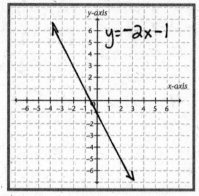

4. a. $y = 4$
   b.

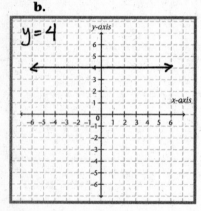

5. a. $y = \frac{2}{3}x$
   b.

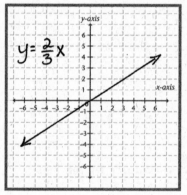

6. a. $y = -\frac{1}{4}x - 1$ or
      $y = -0.25x - 1$
   b.

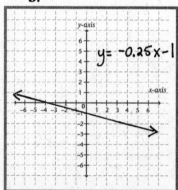

Index

Negative numbers and signs (*cont.*)
  commutative property for, 30
  constants and coefficients,
    95–96
  disguised as subtraction,
    137–139, 141
  distributing, 136–137
  exponents and, 244–248
  multiplying and dividing,
    41–49, 110–113, 117
  subtracting, 11–12
Negative slope, 291–293
Newhart, Eve, 165
No slope, 300
Notes, rewriting, 305
Number lines
  combining integers on, 5, 6
  fractions and decimals on, 13,
    279, 316
  solution sets for inequalities on,
    210–223
Numerical coefficients, 89

Opposites, 40–42, 49
Order of operations, 20–22, 26,
    34, 35, 56, 112, 133, 175–177,
    181, 182, 189, 251
Ordered pairs, 270, 273, 274,
    280–285, 302
Origin, 279, 282, 283

Panda Rule (*see* PEMDAS)
Parentheses
  using, 7, 9, 24, 133, 252
PEMDAS (Parentheses, Exponents,
    Multiplication, Division,
    Addition, Subtraction) order
    of operations, 20–22, 26, 34,
    35, 56, 112, 133, 175–177,
    181, 182, 189, 251
  undoing PEMDAS, 175–179, 189
Percents, 316
Perry, Stephanie, 37–38
Photographers, 165
Planes (*see* coordinate planes)

Points (*see also* ordered pairs)
  on a plane, 279, 280
  plotting/graphing, 281–283
Popularity, 23, 26, 29
Portman, Natalie, 110
Positive integers, 2
Powers, 317
Pretend Cheat Sheet, 275,
    306–308, 310
Procrastination, 226–228, 275

Quadrants, 280, 282, 283
Quiban, Maria, 128–129
Quickie Sheet, 309–311
Quizzes
  Are You a Stress Case?, 77–82
  Do You Pick *Truly* Supportive
    Friends?, 231–236

Rational numbers, 211–212,
    313–315
Ratios, 289–290, 313
Rays, 210
Real numbers, 283, 314
Reciprocals, 42
Relations, 267
Repeated multiplication, 238

Sausage Factories (*see also*
    functions), 265–269, 301
Scholarships, 13
Scratch notes, 306
Sets of numbers, 313–314
Simplifying expressions, 21–22,
    100–101, 105, 106, 112–114,
    117–119, 122–126, 130, 131,
    138–140
Slope, 289–290, 293–297, 302
  negative, 291–292
  slope of zero vs. no slope, 300
Slope-intercept form for lines,
    295–300
Slope of zero, 300
Solution sets, 215, 216, 219–223, 225

About the Author

Best known for her roles as Winnie Cooper on *The Wonder Years* and Elsie Snuffin on *The West Wing*, Danica McKellar is also an internationally recognized mathematician and a two-time *New York Times* bestselling author, for *Math Doesn't Suck* and *Kiss My Math*.

Upon the release of her groundbreaking debut, *Math Doesn't Suck*, Danica made headlines nationwide. Named "Person of the Week" by *ABC World News with Charles Gibson*, her continued success now makes her a regular on shows such as *Good Morning America*, *The Today Show*, and even NPR's *Science Friday*. Danica has also been honored in Britain's esteemed *Journal of Physics* and *The New York Times* for her prior work in mathematics, most notably for her role as coauthor of a mathematical physics theorem that bears her name (the Chayes-McKellar-Winn Theorem).

A summa cum laude graduate of UCLA with a degree in Mathematics, McKellar's passion for promoting girls' math education has earned her multiple invitations to speak before Congress on the importance of women in math and science. Amidst her busy acting schedule, Danica continues to make math education a priority as a featured guest and speaker at mathematics conferences nationwide.

McKellar is also a spokesperson for the Math-a-Thon program at St. Jude Children's Research Hospital, which raises millions of dollars every month both for cancer research and to provide free care for young cancer patients.

McKellar lives in Los Angeles, California. This is her second book.